I0504348

Human Computation

A Transdisciplinary Journal

Volume 1 • Issue 1 • October 2014

http://hcjournal.org

Free for readers, free for authors.

Human Computation is an international and interdisciplinary forum for the electronic publication and print archiving of high-quality scholarly articles in all areas of human computation, which concerns the design or analysis of information processing systems in which humans participate as computational elements.

Publisher:
Human Computation Institute
Fairfax, Virginia
USA

ISBN 978-1502732033
ISSN 2330-8001

Human Computation – Editorial Board

Human Computation

Volume 1, Issue 1   October 2014

Human Computation (2014) 1:1:1-3
© 2014, Michelucci & Simperl. CC-BY-3.0
ISSN: 2330-8001, DOI: 10.15346/hc.v1i1.1

Editorial

PIETRO MICHELUCCI, Human Computation Institute

ELENA SIMPERL, University of Southampton

"In the long history of humankind (and animal kind, too) those who learned to collaborate and improvise most effectively have prevailed." – Charles Darwin, 1859

"The computer is incredibly fast, accurate, and stupid. Man is incredibly slow, inaccurate, and brilliant. The marriage of the two is a force beyond calculation." – Leo Cherne, 1968

---

1. INTRODUCTION

We are not the first to recognize the tremendous potential of human cooperation, nor are we the first to consider the capabilities that might be enabled by bringing together humans and computing machines. Nonetheless, with this inaugural issue of *Human Computation* we come together earnestly, as a global, interdisciplinary scientific community to manifest these phenomena toward their very advancement.

The diversity of our community, however, gives rise to both benefits and challenges. While we can expect to inspire new ideas by sharing discoveries, methods, and results across disciplines, we may also encounter barriers of culture and communication. Indeed, the development of our first few issues has already revealed the complexity of executing peer review in such a space. This experience has reinforced the guiding principle that the content of this journal is driven by purpose rather than approach. We use the term "transdisciplinary" to convey this emphasis.

Thus, we seek to combine perspectives and research within and across scholarly disciplines to further the development of theoretical frameworks, design methods, analytic approaches, interaction techniques, participatory mechanisms, as well as technology and infrastructure toward understanding and implementing information-processing systems that combine humans and machines to achieve unprecedented capabilities.

2. STATE OF THE FIELD

These are exciting times for Human Computation researchers. The publication of this issue occurs on the heels of a Computing Research Association visioning activity to develop a U.S. national research roadmap in human computation, and directly precedes the second annual HCOMP conference, to be held in Pittsburg next month. In parallel to these activities, a burgeoning community of citizen scientists, which also employs and innovates human computation methods, has been self-organizing by creating a new professional association and launching a forthcoming journal of its own. As industry and government interest in human computation grows and public awareness increases, we can only expect to see increasing opportunities for research in this space.

3. CONTENTS OF THIS ISSUE

It is one thing to assert the existence of collective intelligence, but quite another to measure it. Simon DeDeo provides an illuminating and rigorous treatment of collective intentions, beliefs, and values. Using Wikipedia as a case study, he evaluates the hypothesis that group level mental states can be distinct from and irreducible to the mental states of its individual constituents.

The work of Lora Aroyo and Chris Welty shares a similar goal, though they approach the problem of measuring crowd behavior and performance from a slightly different perspective. They propose a revised model for human-derived 'gold standards' conventionally used in Computer Science to assess the accuracy of any information processing task. More specifically, their model explores the tantalizing possibility that disagreement among crowdworkers, rather than merely indicating poor annotation quality, may actually be informative about problematic tasks, and go on to provide an empirical framework that captures these cases and celebrates the diversity of human opinion.

A cornerstone of theoretical Computer Science has been the ability to specify the complexity of computational problems; that is, to classify such problems in terms of the resources needed to solve them. This has enabled computer scientists to understand which problems are best suited to algorithmic (deterministic) processes and to align a given problem with the most appropriate approach. R. Jordan Crouser, Benjamin Hescott, and Remco Chang boldly undertake the challenge of defining an analogous complexity framework for systems with humans-in-the-loop, which begins to build a theoretical foundation for performance-based modeling in human computation.

Until recently, crowdsourcing has been an offline task due to the serialized process of first recruiting participants and then collecting data. Walter Lasecki, Christopher Homan, and Jeffrey Bigham survey new, on-demand crowd-powered systems in order to provide an architectural framework, replete with techniques and careful instructions for replicating recent successes in designing and implementing such real-time systems.

Celebrated cultural anthropologist Mary Catherine Bateson concludes this issue with an incisive perspective that considers the interplay of individuality and connectedness toward self-actualization. As we build systems that collect and aggregate human contributions, we are cautioned to preserve and recognize the impact of individual voices and actions.

We hope you find this first issue of *Human Computation* both stimulating and useful. We are grateful to our initial contributors for choosing this forum for their high quality work and to our esteemed board of editors and advisors for their enduring support as we worked together to build and launch this journal. And finally, dear reader, we look forward to including your own scholarly work in future issues.

Collectively yours,

Pietro Michelucci & Elena Simperl
Co-Editors-in-Chief
Human Computation

Human Computation (2014) 1:1:5-29
ISSN: 2330-8001, DOI: 10.15346/hc.v1i1.2

Group Minds and the Case of Wikipedia

SIMON DEDEO, SCHOOL OF INFORMATICS AND COMPUTING, INDIANA UNIVERSITY, BLOOM-

INGTON, IN 47408 & SANTA FE INSTITUTE, SANTA FE, NM 87501

ABSTRACT

Group-level cognitive states are widely observed in human social systems, but their discussion is often ruled out *a priori* in quantitative approaches. In this paper, we show how reference to the irreducible mental states and psychological dynamics of a group is necessary to make sense of large scale social phenomena. We introduce the problem of mental boundaries by reference to a classic problem in the evolution of cooperation. We then provide an explicit quantitative example drawn from ongoing work on cooperation and conflict among Wikipedia editors, showing how some, but not all, effects of individual experience persist in the aggregate. We show the limitations of methodological individualism, and the substantial benefits that come from being able to refer to collective intentions, and attributions of cognitive states of the form "what the group believes" and "what the group values."

1. INTRODUCTION

Accounts of the history and structure of human behavior naturally attribute cognitive properties not just to individuals, but to groups. They do so despite strong methodological rules against such talk (Arrow, 1994): fields as varied as political science, history, anthropology and economics make constant reference to the attitudes, expectations, beliefs, and values of groups. Scholars talk about the desires and strategies of a social class (Kołakowski, 2008), the expectations of financial markets (Kirman, 2010), or the attitudes of a nation (Elias, 2000a). It is generally—usually implicitly—understood that talk about group-level mental states is not simple shorthand for talk about the mental states of individuals. The relationship between (for example) "society's values" and the behaviors and beliefs of its citizens, or "the beliefs of the committee" and those of its members, is still at best only partially understood.

Reference to group-level cognitive properties in the social sciences has thus been contentious (Wegner, 1987). Yet the idea of minds overlapping and nesting comes naturally to those in artificial intelligence, where mind is understood solely as computation. As far back as 1967, Marvin Minsky wrote that "the question of where a particular machine ends and its environment begins can be settled only by a convention or definition" (Minsky (1967), Pg. 19) and urged that methodological criteria alone settle the question of where boundaries should be drawn.

The very success of the traditional psychological sciences points to the empirical utility of the biological boundary. In this article we provide a scientific account of how that boundary can be violated in informal, self-organizing social systems.

We shall show how reference to the irreducible mental states, and psychological dynamics, of a group mind is necessary to build sensible mathematical theories of large scale social phenomena. To do so, we provide an explicit example drawn from recent work on cooperation and conflict among Wikipedia editors (DeDeo, 2013).

Section 2, "Joint Machines and Social Variables", introduces our account of group minds by reference to finite state machines. We show how the desire to describe coarse-grained properties of interacting machines leads naturally to the use of computational states that are disjunctive unions of the underlying mental states of the individuals involved.

Section 3, "Group Minds in Wikipedia," provides an empirical example of how the considerations suggested by the toy models of Section 2 play out the empirical dynamics of cooperation and conflict on Wikipedia. We build intuition for the problem by reference to a finite-state model of behavior among a large group of heterogenous agents, and find the Nash Equilibrium explicitly. We then show how reference to the properties of group minds is essential for a parsimonious account of real-world behavior.

Section 4, "Subpopulations and Subminds," shows how, in Wikipedia, features of an individual's experience may, or may not, alter group-level accounts. Wikipedia users can interact in a variety of ways, including via discussion on article "talk" pages. Some forms of interaction leave strong traces at higher levels of organization, while others—including talk page interaction—coarse-grain away.

Section 5, "Conclusions", places these results in a larger context, connecting our work to recent work in neuroscience, complex systems, and political science.

In two Appendices, we provide additional detail on our statistical methods, and on an interesting subtlety in the use of reverts to study conflict on Wiki-like systems.

2. JOINT MACHINES AND SOCIAL VARIABLES

Finite-state machines provide our first example of the transition from individual to group-level accounts of cognition.

Such machines model individuals as cognitively-limited transducers: conditional upon an environmental input and the current internal state, the individual produces a behavioral output and changes state. As the name suggests, the individual has only a finite number of possible internal states, and thus an explicitly bounded number of response-patterns. Such machines provide a highly simplified account of cognition, with internal states associated with a mental state or "disposition".

The famous "tit for tat" (T4T) solution for iterated prisoner's dilemma is a particularly simple example. Written as a finite-state machine, T4T has two internal states, C and D, which both name the internal state and dictate the machine's next action, cooperate or defect. Depending on the action previously observed, it shifts between these two states (Fig. 1a). An opponent's defection (d signal) in the current round shifts the machine to state D, and it will defect in the next round.

2.1. **Joint Machines and Computational Macrostates**

The T4T solution gained early attention in the quantitative study of altruism and social cooperation. It did so both because of its natural interpretation in terms of (stylized) facts about social conventions, and because of its behavior under noise: T4T is "forgiving", and allows opposing strategies to recover from an accidental defect.

This can be seen in Fig. 1(b), where we show the *joint machine* that emerges when two T4T strategies interact on a noisy channel. Each state of Fig. 1(b) corresponds to a joint specification of the underlying computational states of the two individuals. With some probability ε, a transmitted symbol is flipped to its opposite; it is this flipped symbol that is both received by the other machine and determines the payoff in that round. Noise can send a cooperative pair into a cycle of alternating, or even mutual, defection, but with non-zero probability of recovering the cooperative solution.

In the T4T joint machine, the states of the system are also causal states of the fine-grained system (Crutchfield and Young, 1989); knowledge of the current play (*CC* vs. *CD*, and so forth) tells us everything we need to know about the future probabilities of action.

When it comes to building accounts of social systems, we are necessarily concerned with aggregate properties; a theory of social systems is not an account of the complete psychological states of the individuals, but is rather phrased in terms of social variables that summarize the underlying states. As an example, we may be interested not in the dynamics of the individual psychological states of the T4T machines, but only in the (absolute) inequality of outcome. In prisoner's dilemma, the absolute difference in payoff is either zero (for *CC* and *DD*—both "win" or both lose) or Δ (greater than zero; in the case where one player receives the payoff $w > 0$, and the other player receives the "sucker's payoff", $s < 0$).

These coarse-grained equivalence classes define *macrostates* (Shalizi and Moore, 2003) of a higher-level theory; the dynamics of these macrostates are effective theories (DeDeo, 2011). In the joint-T4T case, the macrostates are, again, causal states: knowledge of the current payoff inequality tells the observer everything she needs to know about the future behavior of the system. (Macrostates need not be causal states, as can be seen in for a $0/\Delta$ coarse-graining in the interaction of two "tit-for-two-tats" (T42T) strategies, where T42T is defined as "defect if and only if defected against twice in a row.")

A coarse-graining that refers unambiguously to the state of one, and only one, of the two individuals—*e.g.*, a coarse-graining to *CC*-or-*CD* and *DC*-or-*DD*—does *not* provide causal states and thus can, at best, provide only imperfect predictions of future behavior. By contrast, it is easy to construct situations in which the $0/\Delta$ macrostates—and only these macrostates—are relevant to a third-party observer. For example, one can augment the prisoner's dilemma with a third player who "wins" if she correctly picks the inequality level. In this case, the third player never has cause to refer to states of the individual players; she need only refer to disjunctive unions (*CC*-or-*DD* vs. *CD*-or-*DC*).

The desire to talk about coarse-grained properties of a system leads us to theories whose basic units are disjunctive unions of joint mental states. In this simple case, there is no need to refer to the properties of individuals; not only is our theory simpler when we work at the group-level, we lose no predictive power.

2.2. **From Machine to Mind**

The previous section provided a toy example of the Minsky intuition: that computational accounts of behavior may remain agnostic about the boundaries between interacting systems. We have further shown how the social variables relevant to a third agent may mean that the relative computational states of the system do not uniquely and separately identify the mental states of individuals.

Finite-state accounts, however, leave much to be desired. They provide no real space for intentionality, or the "aboutness" of internal states; it is only by strained analogy, for example, that one can claim a finite-state machine believes something about its environment.

In addition, our use of only two (or three) individuals allows us to demonstrate only a very limited gain from reference to joint mental states. Let us take an informal example. One might refer to a store as "friendly" if at least one of its two clerks decides to help. But such an attribution, of a mental state (willingness to help) to a group agent ("the store"), provides little gain over and above the statement that "at least one of the two clerks was willing to help", and this latter statement makes no reference to mental properties other than those of the individual agents.

Finally, it is hard to see in these simple examples how someone could come to know the macrostate without coming to know the microstate. The reason I call a store friendly is precisely because I've observed a "normal" mental property of one of the clerks, and similar objections would appear to apply in any natural account of coming to know the macrostates of Fig. 1(c).

When both the number of individuals and the space of possible behaviors is small, in other words, talk of group minds may be easily schematized—but provides only limited benefit. In the next section, we will see how they become essential in the description of large-scale social phenomena.

3. **GROUP MINDS IN WIKIPEDIA**

In this section, we will demonstrate that observed behavior on Wikipedia urges us to postulate the existence of group-level cognitive states and laws.

We first present a game-theoretic analysis of conflict on Wikipedia. We then show how the observed phenomenology requires us to modify and extend the simple assumptions of this game. We show how these extensions lead us to an account of group cognition that is not simply reducible to that of the individuals involved.

3.1. **The Wikipedia System**

Founded in 2001, the online encyclopædia Wikipedia has been a source of scholarly attention for over a decade (Bar-Ilan and Aharony, 2014). While the legally-recognized Wikimedia Foundation has on the order of 100 employees, day-to-day behavior of the 10^5 volunteer editors is guided by a set of overlapping policies, guidelines, permanent and *ad hoc* committees, formal and informal rules, a range of behavioral norms and cultural practices, and (perhaps most importantly) the pragmatic expectations pseudonymous users form of each other.

Such a large population ends up a cohesive, though far from conflict-free, culture that has been the source of ethnographic studies (Reagle, 2010; Jemielniak, 2014), interventional experiments (*e.g.*, Kiesler et al. (2012); Towne et al. (2013)) and quantitative observational studies (*e.g.*, Lam and Riedl

(2009); Welser et al. (2011); Greenstein and Zhu (2012)). From around 2006 to 2012, the number of "active" volunteer editors ranged between 20,000 and 60,000; the plateau and then decline in the number of users identified by Halfaker et al. (2012) points to the long-term evolution of a project that six years previously underwent rapid growth.

3.2.   Social Norms, Page Reverts, and a Square-Root Law

To study the characteristics of this system, we focus on the page-by-page time-series of "reverts" in the encyclopedia. Reverts—informally, when one editor "undoes" the work of another—are an excellent tracer of short-term conflict (Kittur and Kraut, 2010); they also play a central role in the social norms of the larger Wikipedia community, where users discuss "Edit Warring" (repeated undoing of other's work) in depth, and the "three revert rule" (do not revert more than three times on a single page) makes direct reference to the behavior as a bright line for policy violation.[1] Though necessarily imperfect, automated identification of reverts allows us to track sites of conflict over $\sim 10^6$ individual actions by $\sim 10^5$ editors,[2] spread over a set of 62 highly-edited pages.

We focus on the statistics of "continued non-reversion" which we refer to informally as "cooperative runs". In particular, we study how the probability that the next edit will be a revert declines as a function of time since the last revert. The phenomenological form of this law was found in DeDeo (2013) to be

$$P(R|RC^k) = \frac{p}{(k+1)^\alpha},\qquad(1)$$

where $P(R|RC^k)$ is the probability that one sees a revert edit after exactly-k non-reverting edits, and p and α are constants. Empirically, α ranges between 0.4 and 0.8, with an average (in our current data-set) of 0.5 ± 0.1, an approximate inverse square-root scaling of propensity to revert with length of cooperative run.

We are concerned only with how that small section of history, from the recent revert to the current edit, affects behavior; explicitly, our measurements average out over variety of larger contexts that

---

[1] See https://en.wikipedia.org/wiki/Wikipedia:Edit_warring (last accessed 3 July 2014). For highly-edited pages, the main exception to non-reverting as a social norm is the case of "obvious vandalism": editors are encouraged to revert contributions that vandalize a page, where vandalism is defined as a "deliberate attempt to compromise the integrity of Wikipedia. Examples of typical vandalism are adding irrelevant obscenities and crude humor to a page, illegitimately blanking pages, and inserting obvious nonsense into a page". See https://en.wikipedia.org/wiki/Wikipedia:Vandalism, last accessed 6 July 2014. How the community handles vandalism provides a source of background noise to our study—in particular, we are "off by one" in counting the position of the norm-violating action. Vandalism levels are influenced by "page protection", a top-down action that can restrict the editing of articles to different user classes. Early data on protection was not systematically logged, but from 23 December 2004 onwards, Wikipedia software maintained records via version 1.4 of the MediaWiki software (see https://en.wikipedia.org/wiki/Wikipedia:Village_pump_(technical)/Archive_104, last accessed 6 July 2014). The protection status of the George_W._Bush page changed 243 times between then and 3 July 2014; over the same period, the page saw 6,339 non-reverting runs. At most, therefore, less than 4% of non-reverting runs had continuation probabilities influenced by changes in top-down regulation.

[2] For the 979,426 editing actions in our set, we have 224,080 "users"; despite Wikipedia norms that encourage the use of a single, named account, 140,196 are identified only by IP address, and 83,884 by username.

change on longer timescales. This allows us to build up sufficient statistical power to study the detailed structure of the conditional behavior described by Eq. (1).

The timescales of these runs are indeed short. Fig. 2 shows the cumulative distribution of timescales of cooperation for a sample of $171,107$ cooperative runs measured over 62 highly-edited pages. Although occasional long-tail events are possible, more than 80% of the time, a page in our sample can not go more than a day without a revert. Only a small number of cooperative runs last longer than a week; many of these observations come from early in the encyclopedia's history, and concern less controversial topics. On the other extreme, a small fraction of cooperative runs (2.5%) are over (*i.e.*, terminated by a revert) in less than a minute.

Cooperative events are not only reasonably rapid, but also highly social. Fig. 3 shows the distribution of the number of users involved in these runs. For $k > 1$, over 85% of all runs involve at least three users, and the longest runs can draw in dozens. Interactions over and above the commonly-studied dyadic, or pair-wise, case are essential to the phenomenology we observe.

This sociality involves users who may engage for only short periods. Fig. 4 shows the cumulative distribution of "residence times"—the elapsed time between the first and last edit made by a particular user on a particular page. While the timescale of user engagement is broad, spanning nearly eight orders of magnitude, more than half of users have a total engagement time of less than a day.

We will return to these three facts—concerning the rapidity, sociality, and turnover of the community—when we come to modify and extend the simple game-theoretic account we present in the following section.

3.3.   **The Assistance Stage Game for Wikipedia**

This section presents a game theoretic model of revert behavior on Wikipedia. Such a rational-choice account will provide a guide for how to map, in a rigorous fashion, cognitive states—beliefs about the world—to actions. As we shall see, these standard models fall short of describing real-world behavior: they require us to believe implausible facts about individuals.

We model the decision problem faced by users as a game of perfect, but incomplete, information (Kreps, 1990). Our model is emphatically *minimal*: rather than propose a full account of the motives and contexts for individual editors, we construct a parsimonious account consistent with the most basic facts about interaction on Wikipedia.

In particular, we introduce three important facts that constitute distinctive features of Wikipedia-like systems.

Firstly: no top-down agency chooses which user next edits a page. The order in which users edit, and the identity of who will edit next, is uncertain. In this first analysis we approximate the arrival of users as an IID process.

Secondly: the page history is visible to users; individuals are aware when those who come before have chosen to revert. This defines the information available to users.

Thirdly: individuals have only incomplete information about the motives and desires of their fellow users. We model this incompleteness as an uncertainty about whether the next user will perceive the revert/don't-revert choice as (on the one hand) a problem of misaligned incentives or (on the other) a problem of aligned incentives and mutualism.

While these facts capture the sources of knowledge and uncertainty most relevant to system function, they are only imperfectly observed in the real world. In Sec. 4, below, we critically evaluate the relaxation of the third, and strongest, assumption.

For now, given these assumptions, we construct an extensive-form stage game with a randomly-selected individual making a single choice: to revert (R), or cooperate (C). Her payoff for this move is the sum of two games, one played with the individual who edited just-previously, and the other played with the next editor to appear. The payoff matrix of each game is of the form

$$
\begin{array}{c|cc}
 & C & R \\
\hline
C & c+\varepsilon\theta_1, c+\varepsilon\theta_2 & s+\varepsilon\theta_1, w \\
R & w, s+\varepsilon\theta_2 & 0,0
\end{array}
\tag{2}
$$

where ε is a constant, $0 < c < w$, and $s < 0$. We model a heterogeneous population of players using θ_1 and θ_2, independent random variables drawn uniformly between zero and unity. These are chosen once (and for all time) for each player, and boost the payoff associated with the normative action.

As an example, say that the player in question, Alice, chooses to not to revert; say further that the previous player, Walrus, also chose not to revert; say finally that the player after Alice, Carpenter, responds to Alice by choosing to revert. Alice's payoff is $(c+\varepsilon\theta_1) + (s+\varepsilon\theta_1)$. This is her CC payoff from cooperating with Walrus, plus the negative "sucker's payoff", s, from being reverted by Carpenter. On making her move, Alice is aware of Walrus's choice, but does not know the identity (or move choice) of the next player.

Depending on the value of ε and θ_1, Alice's payoff might be positive (because of her cooperation with Walrus, and a general desire to follow social norms against reverting), or negative. Meanwhile, we do not know what Carpenter's final payoff will be until the player after him makes a move, but we know that it is either w (if the next person reverts) or $2w$ (if the next person cooperates).

The random variable governs the uncertainty a user has about her fellow-user's perception of the game; to be clear, Alice knows her θ value (and thus whether she likes to revert others, or prefers to find a cooperative solution), but does not know the character of the next editor. When formulating her strategies for play, therefore, she will have to estimate her utility in the face of uncertain knowledge of others.

3.4.   What Alice Believes about Walrus and the Carpenter

We can make Alice as intelligent as we like, but what sets the complexity of her behavior are the beliefs she holds about the people she encounters.

We shall (at first) give Alice uncharitable views of the sophistication of her fellow editors. In particular, we shall have her believe that editors condition their behavior solely on whether the previous editor did, or did not, revert. She has, in other words, two beliefs: β_c (the probability that the next editor will revert her, given that she cooperated) and β_r (the probability that the next editor will revert her, given that she reverted).

Let us say the previous editor, Walrus, cooperated. Then Alice's best response (she will reason) is to cooperate if and only if

$$(c + \varepsilon\theta_1) + [(c + \varepsilon\theta_1)(1 - \beta_c) + (s + \varepsilon\theta_1)\beta_c)] > w + w(1 - \beta_r). \tag{3}$$

This inequality should be easy to read; the first term (on the LHS) is the benefit to cooperating with Walrus; the second term (in square brackets) is the expected benefit to cooperating with the next editor, given expectations about the next editor's strategies. The third and fourth terms (on the RHS) are (respectively) the (now known) payoff from reverting Walrus, and the expected gain given the possibility that the next user will cooperate (with probability $1 - \beta_r$).

Conversely, and by similar reasoning, if Walrus reverted, Alice's best response is cooperation if and only if

$$(s + \varepsilon\theta_1) + [(c + \varepsilon\theta_1)(1 - \beta_c) + (s + \varepsilon\theta_1)\beta_c)] > w(1 - \beta_r). \tag{4}$$

If Alice has no way of predicting the identity of the next editor, and (furthermore) if she truly believes her fellow editors can hold no more sophisticated a memory than C-vs-R—among other things, for example, if she believes that editors do not respond to, or cannot perceive, "punishment"— then she might as well choose the action that most benefits her. In other words, Alice herself will adopt the strategy "cooperate after cooperation" if and only if her θ_1 is such that the inequality of Eq. (3) is satisfied; similarly, she will cooperate with a reverter only if the inequality of Eq. (4) is satisfied.

The final stage of the argument is to find the fixed point where beliefs about the system lead to actions that maintain the truth of those beliefs.[3] If we ask the population as a whole to reason as Alice does, then the probability that the inequality Eq. (3) holds is just the probability that θ for the individual in question is large enough, in particular that

$$P\left(\theta_1 > \frac{2(w - c) + \beta_c(c - s) - \beta_r w}{2\varepsilon}\right), \tag{5}$$

and for beliefs to be consistent with the actions observed, this probability has to equal $1 - \beta_c$. A similar argument applies to the satisfaction of Eq. (4), and, with two equations and two unknowns, we can solve for the self-consistent solution to the beliefs β_c and β_r. The general solution is the ratio of polynomials; for the particular case where the disgruntlement, s, on being reverted is equal to $c - w$, $w - c$ is greater than zero and less than ε, and β_c and β_r are equal, the expected behavior is particularly simple,

$$P(R|C) = P(R|R) = P(R|RC^k) = \frac{w - c}{\varepsilon}, \tag{6}$$

where $(w - c)/\varepsilon$ is a constant.

We emphasize that Eq. (6) is *clearly incompatible* with the observed phenomenology, Eq. (1). This is to be expected: the assistance stage game makes assumptions that we believe are naturally violated in the real world.

---

[3] Game Theorists will recognize this as a Markov Perfect Equilibrium (Fudenberg and Tirole, 1991).

However, Eq. (6) does give us a clear account of how we can alter the model by hand. All we have to do is make ε a function of k—indeed, a very simple function of k, $\varepsilon(k) \propto (k+1)^{\alpha}$—and we can recover Eq. (1) instantly. Such systematic shifts in ε can not be accounted for by the stage game we present above; to explain them we must violate some or all of the assumptions above.

In the next section, we discuss exactly what this violation means for our understanding of cooperation on Wikipedia, and the ways in which we describe the nature of belief and action in a large-scale social system.

3.5.   **Three Theses for a Group Mind**

Changing ε amounts to doing two things simultaneously. First, it amounts to changing the overall balance in the system: shifting agents away from seeing Wikipedia as a series of one-shot prisoner's dilemma games, and towards a mutualistic viewpoint that sees cooperation, even at risk of revert, as intrinsically good. Second, and equally importantly, it amounts to changing the beliefs that individuals have about others. This simple argument, based on long-understood features of game-theoretic models, suggests the first thesis of this section.

> **Thesis one**. *For behavior to shift in the fashion we observe, both beliefs and desires must change in a systematic fashion.*

Belief change with k will naturally occur in the real world. We expect that real Wikipedia users are playing sophisticated strategies, with potentially greater look-back times and sensitivity to features of the system such as the usernames, talk pages, and user pages of the editors currently active. Over the minutes and hours that elapse during the unbroken non-reverting sessions we examine here, they will naturally—and in violation of the simple assumptions of the stage game—come to change their future expectations.

We also expect preference change among Wikipedia users. Such change may be endogenously driven—I may simply enjoy cooperating more when others around me appear to be cooperating— or may be due to complicated mechanisms involving reputation effects that we know to be in effect in other public goods games.

Both beliefs and preferences will be driven by the changing state of the page itself. Pages will improve (or degenerate) over time, and the evolving state of the page will drive users (or their programmers, in the case of bots) towards (or away) from cooperation. The IID assumption will fail as changing conditions shift the composition of the pool of active editors.

The rapid formation of these cooperative runs (Fig. 2) suggests that these effects are unlikely to arise from a top-down process; the involvement of multiple users (Fig. 3) allows for multiple distinct mechanisms for belief and preference change, and a variety of outcomes. The timescales of community turn-over for any particular page, shown in Fig. 4, further suggests that users do not have sufficient time to model each other one-by-one. Both theoretical and empirical considerations, therefore, suggest we adopt a second thesis:

> **Thesis two**. *The cognitive laws we can infer from Eq. (1) do not simply reduce to direct accounts of mental states of individual editors.*

Sec. 3.4 showed us how to infer beliefs and preferences from actions; but both mechanistic and theoretical considerations make it unlikely that these beliefs ("the next person will revert with probability proportional to $\varepsilon(k)$") and preferences ("my desire to follow social norms experiences an increase proportional to $\varepsilon(k)$") correspond to the beliefs or desires of any user whatsoever.

In other words: Alice, Walrus and the Carpenter may be useful explanatory devices, but they do not correspond to any particular users in the real world. Real users are more limited (in the extent to which they can gather information about behavioral patterns), more nuanced in their desires, and likely attribute greater cognitive complexity to each other. However, knowledge and simulation of this reality is unnecessary to describe the behavior we observe.

Together, these two theses—a positive thesis connecting behavior to cognitive state, and a negative thesis preventing direct identification of those cognitive states with those of any individual—urge us to accept a third thesis, and the main result of our paper:

> **Thesis three.** *The reliability of Eq. (1) should be explained by the existence of psychological laws pertaining to group-level mental states.*

Different disciplines have different traditions for talking about mental states and laws of this form. Economics, as mentioned above, has little difficulty referring to the beliefs of "the market" without necessarily attributing these beliefs to any participant (Hayek, 1945). Political scientists study core features of civil conflict by reference to group-level values and ideologies, even (and, indeed, particularly) when these conflict with the beliefs of group members (Sanín and Wood, 2014). Other fields, such as intellectual history, may talk about such states by reference to an idealized individual, whose cognitive features are understood by comparison to biographical accounts of actual thinkers (Hazard, 2013).

4.   SUBPOPULATIONS AND SUBMINDS

Wikipedia norms urge that editor-editor communication take place on the associated talk page of the article in question;[4] meanwhile, users can, and sometimes do, make edits to a page numerous times in a short period. Both of these practices provide information to others about an editor's goals, desires, and level of cooperativity over and above that possible in the stage game above.

To see if these micro-level facts "bubble up" to alter large-scale features of behavior, we can look at how behavior shifts within the subgroup of users who have recently interacted. To do this, we can look at the conditional behavior of this group, using the same phenomenological law of Eq. (1). Taking "recent" interaction to mean "within the last ten edits on the page," we can define

$$P(R|RC^k, \text{int}) = \frac{p_{\text{int}}}{(k+1)^{\alpha_{\text{int}}}} \qquad (7)$$

---

[4]See http://en.wikipedia.org/wiki/Wikipedia:Talk_page_guidelines ("The purpose of a Wikipedia talk page ... is to provide space for editors to discuss changes to its associated article or project page") and http://en.wikipedia.org/wiki/Wikipedia:Etiquette ("When reverting other people's edits, give a rationale for the revert (on the article's talk page, if necessary), and be prepared to enter into an extended discussion over the edits in question"). Both pages last accessed 12 September 2014.

as the probability that, when a user who has edited the page within the last ten edits performs a *new* edit, that edit is a revert. If the parameters that define $P(R|RC^k, \text{int})$ and $P(R|RC^k)$ differ, this is a signal that the effective mental states of these subgroups evolve differently.

Fig. 5 shows, for the 62 pages in our data set, how α and $P(C|RC)$ shift when one considers the population as a whole as compared to two sub-populations: editors who have recently (within the last ten edits) made an edit, and those who have not.

Parameters for the two sub-groups show a clear separation. When restricting to those editors who have recently interacted, one finds a similar $P(C|RC)$—*i.e.*, a similar disposition at the beginning of a cooperative streak—but a much larger α. Instead of the square-root law, we find instead that the probability of a revert declines roughly as $1/k$. Informally, group-level dispositions for editors who interact begin similarly, but shift more rapidly towards cooperation. We emphasize that this "learning" describes a process that occurs at the group level—any individual editor has a complex set of beliefs and desires, but their effects can be explained by reference to a collective mental state of the population of which she is a member.

We also consider the effect of "Talk Pages"—associated pages explicitly devoted to the discussion of contentious issues in the page itself. Fig. 6 shows, for 62 pages in our data set, how α and $P(C|RC)$ shift when one considers the population as a whole, and two sub-populations: of editors who have recently (within the last ten edits) made an edit to the corresponding talk page, and those who have not.

Surprisingly, and in contrast to the findings for on-page interaction, we see little difference in behavior between the group of those who have, and those who have not, recently used the talk page. A similar result applies when we extend this window further to include talk page interaction within the past one-hundred edits. Talk does not shift behavior: though we emphasize, in keeping with our focus on group-level dynamics, that this conclusion applies to the joint state of the "talk page users" population, and not to any particular user.

Indeed, it is likely that individual users shift their opinions and beliefs a great deal through the unstructured discussion possible on talk pages. However, these shifts appear not to add together in a coherent fashion and do not leave a strong trace at higher, and more coarse-grained, levels of organization. What shifts preferences and beliefs at the group level is not talk, it seems, but action.

5. CONCLUSION

This paper has worked through a minimal model of interaction on a large-scale networked social system. We showed how game-theoretic notions provided us a rigorous framework within which to move from observations of behavior to knowledge about beliefs and other mental states. The complex and networked structure of the system strongly suggests that we attribute these states to group-level cognitive processes.

Our basic model for interaction locates the struggle for sociality on Wikipedia in the extent to which users perceive the problem as a one-shot prisoner's dilemma, as opposed to a mutualistic interaction where the greatest benefits accrue to joint norm-following. This places our account of Wikipedia on the continuum between those who identify the origin of social complexity in altruism and group selection, vs. those who see the burden on the side of cognitive ability (see, *e.g.*, extended discussion in Tomasello (2009); Bowles and Gintis (2011)).

The recursive nature of both problems means that it not only the perception of users that matters, but their beliefs about the perceptions of others, and so forth. The fixed point solution we describe in Sec. 3.3 implicitly assumes the existence of common knowledge. Real-world studies show that, at best, individuals only approximate such solutions (Frey and Goldstone, 2013); indeed, results of this form, as well as theoretical arguments against the existence of true common knowledge (Gintis, 2014), strengthen our thesis two above—not only will individuals have more complex beliefs and shifting desires than the group-level beliefs, but they will reason in some bounded-rational fashion when deciding to act.

That group-level cognitive states appear to obey reasonably simple laws means not only that outside, scientific observers can form parsimonious descriptions of the system. It also means that such knowledge is potentially available to individuals within the system. Reference to "how cooperative the group is" will likely form part of an individual user's reasoning. Such forms of group-level-pattern recognition are known to play a role in both human (Gigerenzer and Todd, 1999) and animal sociality (Daniels et al., 2012).

As shown in Sec. 4, minds may be nested within minds. Discovery of such subgroup structure parallels the distinct subgroup strategies found in animal sociality (DeDeo et al., 2011), where triadic and higher-level decision-making is likely to be active (DeDeo et al., 2010).

Our results here are complementary to recent work on the quantification of group agency established in a series of influential papers by List and Pettit (List and Pettit, 2002, 2004, 2011). This work establishes that while some limit cases may be trivial (everyone believes X, and so "the group believes X"), theorems on preference aggregation mean that the relationship between individual and group-level mental states is complex. Under List and Pettit's analysis, some groups may fail to be true agents able to form intentional beliefs ("minds," in our language) because of the relations between the individuals that compose these groups. For List and Pettit, the ascription requires that individuals dynamically respond to each other in sufficiently cooperative ways. We have no such constraint, but anticipate that only in groups with sufficiently aligned incentives, or sufficient commitments to norm creation and rule-following, will large-scale cognitive laws be useful sources of knowledge.

Our account is complementary to an influential strand of thought that locates the boundaries of mind—and, in particular, of unitary experience—by reference to the degree of integration between subparts (Tononi et al., 1994, 1998; Oizumi et al., 2014). There is no question in this case that, at least on longer timescales (hours to days), users do not interact frequently enough via the Wikipedia interface to justify a claim that these cognitive states are accompanied by qualitative experience. The interaction between users and distributed, higher-level properties associated with group-level beliefs has important commonalities with current thinking on the emergence of life (Walker and Davies, 2013; Walker et al., 2012; Hoel et al., 2013), and with accounts of multi-level selection in evolutionary systems (Traulsen and Nowak, 2006).

Our work connects directly to philosophical accounts of the Group Mind hypothesis (Theiner, 2008; Theiner et al., 2010; Theiner, 2013), in as much as it takes the cognitive states of these larger, non-biological units seriously. When individuals interact in sufficiently complex ways, the same reasons for including features of the external, physical world as part of a cognitive process (Clark and Chalmers, 1998; Clark, 2008) may apply to the inclusion of the cognitive states of others. These aggregate processes—group minds—will naturally have mental properties that differ from the individual minds out of which they are constructed.

The existence of group-level cognitive states argues for cognitive approaches to the evolution of societies, including the quantification of their inferential and information-theoretic properties (DeDeo et al., 2013), potentially over timescales longer than an individual lifespan (Klingenstein et al., 2014). Our work is thus of direct relevance to a long-standing debate in the social and historical sciences (Elias, 2000b). What is the natural unit of analysis for long-term historical change, or the behavior of large groups? Methodological individualism necessarily reduces all talk of group-level beliefs, including accounts of culture and institutions, to the individual (Bowles, 2009; Bowles and Gintis, 2011). From Sec. 2 we know that it is (of course) not impossible to give an account of group-level mental states (the macrostate) in terms of individual mental states (the microstate)—only that, as we see in Sec. 3.5, it may become extremely unwieldy, requiring us to explain simple laws in terms of an enormous number of fine-tuned processes.

6. ACKNOWLEDGEMENTS

I thank the organizers and attendees of the ASU BEYOND Center February 2014 meeting "Information, Complexity and Life", of the 2014 Santa Fe Institute (SFI) Complex Systems Summer School, and of the 2014 SFI Graduate Workshop in Computational Social Science, for useful discussions, and my two reviewers, Theodore P. Pavlic (Arizona State University) and an anonymous referee, for detailed feedback on this manuscript. I thank the City University of New York's Initiative for the Theoretical Sciences for their hospitality while this work was completed. This work was supported in part by National Science Foundation Grant EF-1137929.

7. APPENDIX: MEASURING α AND P

In DeDeo (2013), the relevant parameters p and α, were estimated from the counts of strings of the form RC^kR, under the assumption that events of the form RC^kR are Poisson-distributed with mean varying as a function of k. Here, we present a simpler method for the estimation of α and p from data. In particular, we estimate p and α from observations of $N(R|RC^k)$; using the binomial distribution, we maximize the posterior

$$\log P(\text{data}|p,\alpha) \;=\; \sum_{i=1}^{\infty} (\log n_i! - \log k_i! - \log(n_i - k_i)!)$$

$$+ k_i \log \left(\frac{p}{(i+1)^\alpha} \right) + (n_i - k_i) \log \left(1 - \frac{p}{(i+1)^\alpha} \right), \tag{8}$$

where n_i is the number of times one observes a string of length RC^i, and k_i is the number of times one observes a string of length RC^iR. In words, each term in the sum is the log-likelihood of a string RC^i being followed by a revert, given the choice of p and α.

Eq. (8) gives slightly different values for p and α than the estimation method of DeDeo (2013). In addition to being easier to compute, it makes fewer assumptions about the process—in particular, it does not require one assume that the number of sequences of the form RC^kR is much less than the total length of the series.

Using Eq. (8), it is simple to estimate more complicated conditional probabilities, as required in Sec. 4. Consider the probability that an editor performs a revert action, given (1) k cooperative events have been seen and (2) the editor has previously interacted with the page,

$$P(R|RC^k, \text{int}) = \frac{p_{\text{int}}}{(k+1)^{\alpha_{\text{int}}}}. \tag{9}$$

One can find the maximum a posteriori values of p_{int} and α_{int} by reference to $n_{i,\text{int}}$ and $k_{i,\text{int}}$, where $n_{i,\text{int}}$ is the number of times a cooperative sequence of length i is followed by a C or R move by an editor who has previously interacted with the page, and $k_{i,\text{int}}$ is the number of times a cooperative sequence of length i is terminated by an R move by an editor who has previously interacted with the page.

8.   APPENDIX: THE DISTANT REVERT PROBLEM

Detecting reverts by looking for the exact replication of page content ("hash-reverts") across two edits is a standard method in research into Wiki-like interactions. However, such a technique may miss an important feature of "undo-like" conflict.[5] If we index all "C" edits by the sum total of content changes they make, then we can index "R"-type edits by the set of "C" edits they undo. For example, consider the following sequence of page contents:

> Version 1, User A
> George Washington was our first president.
>
> Version 2, User B
> George Washington was our first president.
> Famously, he cut down a cherry tree while young.
>
> Version 3, User C
> George Washington was our first president.
> Famously, he cut down a cherry tree while young.
> He warned against the formation of political parties.
>
> Version 4, User D
> George Washington was our first president.
> He warned against the formation of political parties.

In the standard method of tracking revert behavior, this sequence contains no hash-reverts—no complete page text appears twice—and would be recorded as $CCCC$. However, notice that User D, while not undoing the work of User C, does undo the work of User B: she deletes the apocryphal story of Washington cutting down the cherry tree. It is reasonable to imagine tracking this sequence as $C_1C_2C_3R_2$—i.e., the last edit as a reversion of the edit C_2. Depending on the relative rates at which users make edits, and other users disagree with those edits and undo them, the dynamics of distant

---

[5]I am grateful to one of my referees, Theodore P. Pavlic, for drawing this feature of Wiki-like conflict to my attention.

reverts can obey very different laws from those of hash-reverts; indeed, it is possible for distant reverts to be uncorrelated (*i.e.*, follow an exponential distribution with α equal to zero in Eq. (1)) while hash-reverts appear correlated.

Tracking of these distant reverts is non-trivial; it is most easily done by defining an edit as a set of minimum-Levenshtein distance edits, using a tool such as UNIX `diff`. We can then define a distant revert as a new edit whose diff is the exact opposite of a previous edit: all additions become deletions, all deletions become additions, and all changes are reversed. An example of a distant revert is the pair of edits made by users `Wiki alf` and `Joker123192` to the `George_W._Bush` page; on 18 May 2009, `Wiki alf` adds a paragraph; and on 29 May 2009, 32 edits later, `Joker123192` takes it back out.[6]

For the `George_W._Bush` page, inclusion of these single-index reverts increases the number of reverts counted from 14,976 to 15,982; this shifts α from 0.41 ± 0.02 to 0.38 ± 0.02. UNIX `diff` works newline-by-newline, so that if Edit 1 adds a sentence, Edit 2 adds a sentence to the same paragraph, and Edit 3 deletes the sentence added in Edit 1, this will appear as a change, rather than a deletion. One can instead work sentence-by-sentence, to look for potential disagreements about content internal to a paragraph. Doing so finds a total of 16,021 reverts, and (within error) does not change the estimate of α. Note our estimate of α for the `George_W._Bush` differs from that in DeDeo (2013); this is both because we include more data (through 4 July 2014), and because we use an improved method for the estimation of p and α.

Strong norms exist against hash-reverts. For example, the idea that a revert leads to the repetition of prior page content is often part of the definition of revert on Wikipedia itself.[7] By contrast, Wikipedia norms are more ambiguous when it comes to distant reverts, and these reverts may be weaker signals of conflict. For intuition as to why, consider the following sequence:

Version 1, User A
```
George Washington was our first president.
```

Version 2, User B
```
George Washington was our first president.
He was against political polarization.
```

Version 3, User C
```
George Washington was our first president.
He was against political polarization.
He warned against the formation of political parties.
```

Version 4, User D

---

[6] See https://en.wikipedia.org/w/index.php?title=Main_Page&diff=290774306&oldid=290770322 and https://en.wikipedia.org/w/index.php?title=Main_Page&diff=293157787&oldid=292576084, last accessed 8 September 2014.

[7] "Reverting means undoing or otherwise negating the effects of one or more edits, which results in the page being restored to a previous version." (see http://en.wikipedia.org/wiki/Help:Reverting). A weaker claim appears at http://en.wikipedia.org/wiki/Wikipedia:Reverting: "Reverting means reversing a prior edit, which typically results in the article being restored to a version that existed sometime previously." Both accessed 8 September 2014.

```
George Washington was our first president.
He warned against the formation of political parties.
```

Here, User D has partially reverted User B, as before; but (it may be inferred) in part because she might believe that User's C's addition renders User B's contribution moot, rather than because she believes User B's content to be unambiguously bad. If User B disagrees with User D, he may roll back to Version 3 or to Version 2; in either case, this will now appear as a hash-revert.

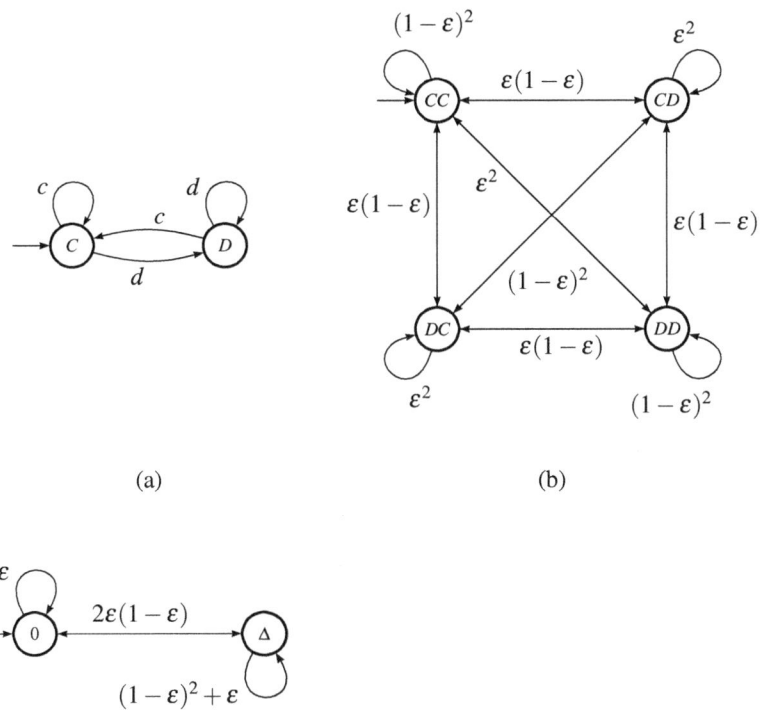

(a)

(b)

(c)

Figure 1. **Machines, joint machines and effective theories.** (a): a finite-state machine representing the "tit for tat" (T4T) strategy in prisoner's dilemma. The node labels the move the agent makes; the arrows label the transitions between states made by the agent upon observing the behavior of its opponent. (b): the joint machine describing the interaction of two T4T agents playing in a noisy environment; nodes label the observed behaviors, while arrows are labelled with transition probabilities. We assume a binary symmetry channel with noise, such that a c (respectively, d) is flipped to a d (respectively, c) with probability ε. (c): a coarse-grained machine describing the evolution of a social variable tracking inequality of outcome; state 0 when both agents receive the same payoff, and state Δ when one receives a higher payoff than the other due to unreciprocated cooperation.

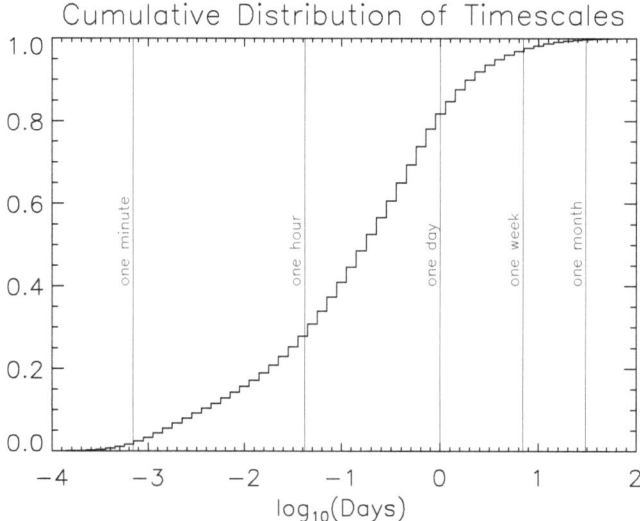

Figure 2. **Cooperation is rapid...** Most cooperative runs last less than twenty-four hours. Plotted is the cumulative distribution of clock-times, for $171,107$ RC^kR runs measured on 62 highly-edited pages, showing that the timescales over which we measure behavioral changes are usually much shorter than day-to-week timescales of exogenous events, and the week-to-year timescales of Wikipedia policy changes.

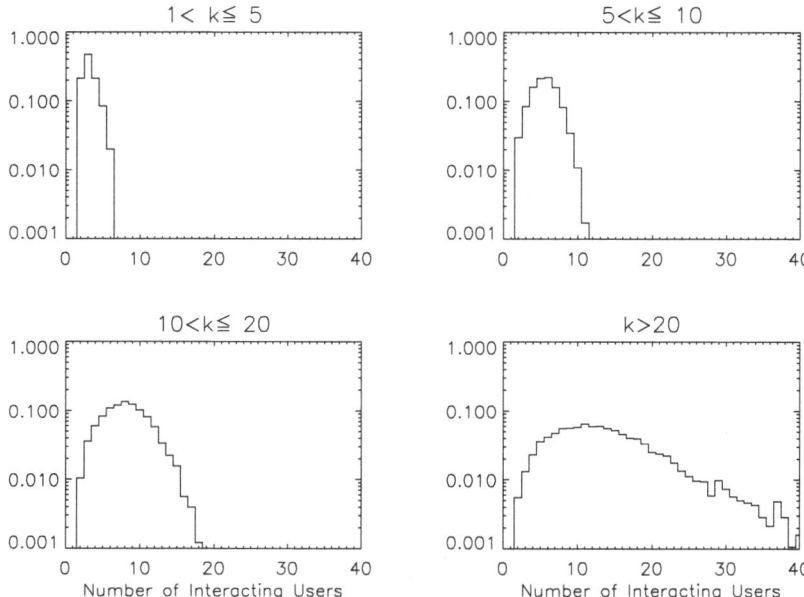

Figure 3. **...and social.** Most cooperative runs involve multi-user interactions. Over 85% of all runs with $k > 1$ involve at least three parties; for longer runs, $RC^k R$ with k greater than five, over 97% of interactions involve at least three parties. The very longest runs, $k > 20$, involve a median of twelve editors. Here we count the unique users responsible for the k C edits and the terminal R, so that (for example) a run of length $k = 2$ can have at most three unique editors.

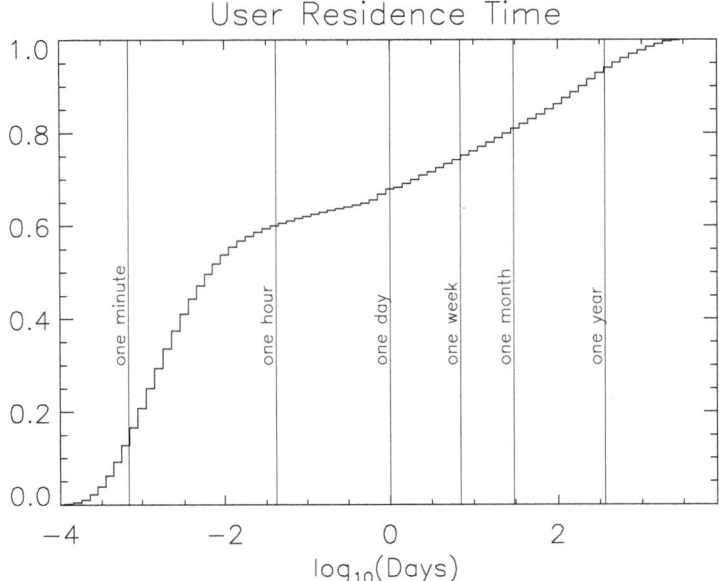

Figure 4. **The majority of users have only short-term engagement with a page.** Plotted here is the cumulative distribution of user longevity—the length of time that elapses between the first and last observed edit on a page—over 62 highly-edited pages. More than half of users have a total engagement time of less than a day; only a small fraction (less than 10%) of users are engaged with editing a page for more than a year.

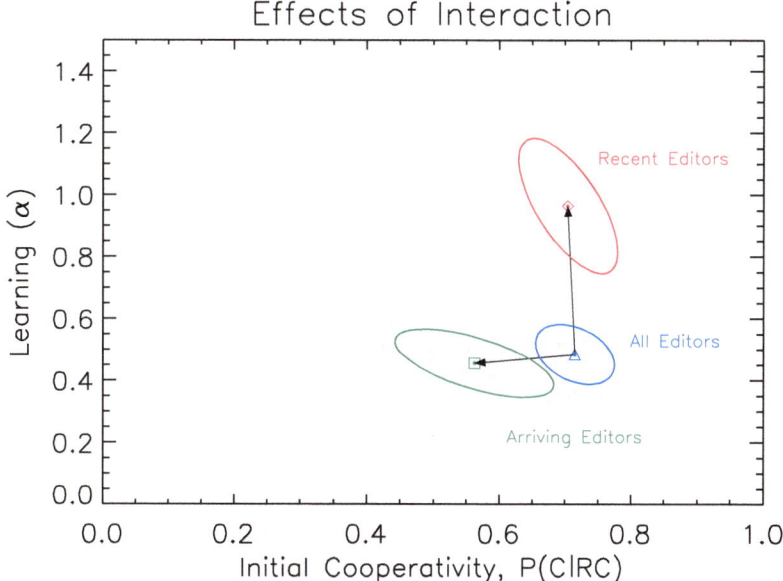

Figure 5. **Users who edit learn faster.** Compared to the population as a whole, users who have edited recently (within the last ten edits) begin with similar predispositions (initial cooperativity, $P(C|RC)$) but shift towards cooperative behavior more rapidly than others. Meanwhile, newly arriving editors tend to be initially less cooperative, and while they do (at the population level) learn to become more cooperative (α greater than zero), they do so more slowly than those who interact. Shown: distribution of p and α parameters for 62 pages in our data, for all users (triangles; blue 1σ range), users who have recently edited (within the last ten edits; diamonds; red 1σ range) and newly-arriving users (squares; green 1σ range).

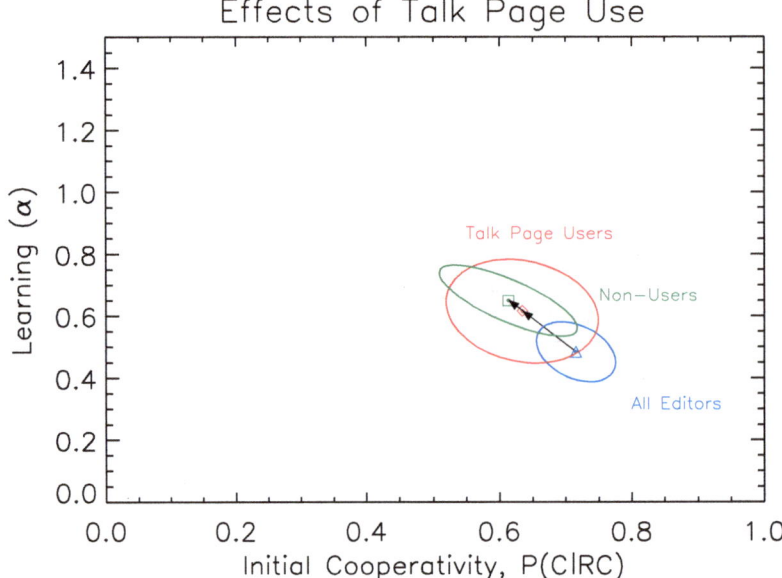

Figure 6. **Talking does not predict cooperation.** Users who have edited the article talk page recently (within the last ten edits on the mainpage) are quite similar to the population as a whole. Shown: distribution of p and α parameters for 62 pages in our data, for all editors (triangles; blue 1σ range), recent talk page users (diamonds; red 1σ range) and non-users (squares; green 1σ range).

9. APPENDIX: WIKIPEDIA PAGES USED IN THIS ANALYSIS

2006_Lebanon_War, Argentina, Atheism, Australia, Barack_Obama, Blackout_(Britney_Spears-_album), Blink-182, Bob_Dylan, Canada, Catholic_Church, Che_Guevara, China, Circumcision, Cuba, Eminem, France, Gaza_War, Genghis_Khan, George_W._Bush, Girls'_Generation, Global-_warming, God, Golf, Heroes_(TV_series), Hilary_Duff, Hillary_Rodham_Clinton, Homosexual-ity, Hurricane_Katrina, IPhone, Iraq_War, Islam, Israel, John_F._Kennedy, John_Kerry, Lindsay-_Lohan, Linux, Lost_(TV_series), Mexico, Michael_Jackson, Neighbours, New_Zealand, Paris-_Hilton, Paul_McCartney, Pink_Floyd, RMS_Titanic, Russo-Georgian_War, Scientology, Septem-ber_11_attacks, Shakira, Star_Trek, Super_Smash_Bros._Brawl, Sweden, The_Dark_Knight_(film), The_Holocaust, Turkey, United_States, Virginia_Tech_massacre, Wikipedia, Wizards_of_Waverly_-Place, World_War_I, World_War_II, Xbox_360

10. REFERENCES

Arrow, K. J. (1994). Methodological Individualism and Social Knowledge. *The American Economic Review* 84, 2 (1994), pp. 1–9.

Bar-Ilan, J and Aharony, N. (2014). Twelve years of Wikipedia research. In *Proceedings of the 2014 ACM conference on Web science.* 243–244.

Bowles, S. (2009). *Microeconomics: Behavior, Institutions, and Evolution.* Princeton University Press.

Bowles, S and Gintis, H. (2011). *A Cooperative Species: Human Reciprocity and its Evolution.* Princeton University Press.

Clark, A. (2008). *Supersizing the Mind: Embodiment, Action, and Cognitive Extension.* Oxford University Press.

Clark, A and Chalmers, D. (1998). The Extended Mind. *Analysis* 58, 1 (1998), 7–19. DOI:http://dx.doi.org/10.1093/analys/58.1.7

Crutchfield, J. P and Young, K. (1989). Inferring statistical complexity. *Physical Review Letters* 63, 2 (1989), 105.

Daniels, B. C, Krakauer, D. C, and Flack, J. C. (2012). Sparse code of conflict in a primate society. *Proceedings of the National Academy of Sciences* 109, 35 (2012), 14259–14264. DOI:http://dx.doi.org/10.1073/pnas.1203021109

DeDeo, S. (2011). Effective theories for circuits and automata. *Chaos* 21, 3 (2011), 037106. DOI:http://dx.doi.org/10.1063/1.3640747

DeDeo, S. (2013). Collective Phenomena and Non-Finite State Computation in a Human Social System. *PLoS one* 8, 10 (2013), e75818.

DeDeo, S, Hawkins, R. X, Klingenstein, S, and Hitchcock, T. (2013). Bootstrap methods for the empirical study of decision-making and information flows in social systems. *Entropy* 15, 6 (2013), 2246–2276.

DeDeo, S, Krakauer, D, and Flack, J. (2010). Inductive game theory and the dynamics of animal conflict. *PLoS computational biology* 6, 5 (2010), e1000782.

DeDeo, S, Krakauer, D, and Flack, J. (2011). Evidence of strategic periodicities in collective conflict dynamics. *Journal of The Royal Society Interface* 8, 62 (2011), 1260–1273.

Elias, N. (2000)a. *The Civilizing Process: Sociogenetic and Psychogenetic Investigations.* Wiley. Second Edition of 1939 Text, edited by Dunning, E., Goudsblom, J., and Mennell, S.

Elias, N. (2000)b. *The Society of Individuals.* Wiley. Edited by Michael Schroter, translated by Edmund Jephcott; first edition, 1987.

Frey, S and Goldstone, R. L. (2013). Cyclic game dynamics driven by iterated reasoning. *PLoS one* 8, 2 (2013), e56416.

Fudenberg, D and Tirole, J. (1991). *Game Theory.* MIT Press.

Gigerenzer, G and Todd, P. M. (1999). *Simple heuristics that make us smart.* Oxford University Press.

Gintis, H. (2014). *The Bounds of Reason: Game Theory and the Unification of the Behavioral Sciences.* Princeton University Press. Second edition.

Greenstein, S and Zhu, F. (2012). Is Wikipedia Biased? *The American Economic Review* 102, 3 (2012), 343–348.

Halfaker, A, Geiger, R. S, Morgan, J. T, and Riedl, J. (2012). The rise and decline of an open collaboration system: How Wikipedia's reaction to popularity is causing its decline. *American Behavioral Scientist* 57, 5 (2012), 665–688. DOI:http://dx.doi.org/10.1177/0002764212469365

Hayek, F. A. (1945). The use of knowledge in society. *The American Economic Review* 35, 4 (1945), 519–530.

Hazard, P. (2013). *The crisis of the European mind, 1680–1715*. New York Review of Books.

Hoel, E. P, Albantakis, L, and Tononi, G. (2013). Quantifying causal emergence shows that macro can beat micro. *Proceedings of the National Academy of Sciences* 110, 49 (2013), 19790–19795.

Jemielniak, D. (2014). *Common Knowledge?: An Ethnography of Wikipedia*. Stanford University Press.

Kiesler, S, Kraut, R, Resnick, P, and Kittur, A. (2012). Regulating Behavior in Online Communities. In *Building Successful Online Communities: Evidence-Based Social Design*, R.E. Kraut, P. Resnick, S. Kiesler, M. Burke, Y. Chen, N. Kittur, J. Konstan, Y. Ren, and J. Riedl (Eds.). MIT Press.

Kirman, A. (2010). *Complex economics: individual and collective rationality*. Routledge.

Kittur, A and Kraut, R. E. (2010). Beyond Wikipedia: coordination and conflict in online production groups. In *Proceedings of the 2010 ACM conference on Computer Supported Cooperative Work*. ACM, 215–224.

Klingenstein, S, Hitchcock, T, and DeDeo, S. (2014). The civilizing process in London's Old Bailey. *Proceedings of the National Academy of Sciences* 111, 26 (2014), 9419–9424.

Kołakowski, L. (2008). *Main Currents of Marxism*. W. W. Norton & Company. Translated by P. S. Falla; first edition, 1976.

Kreps, D. (1990). *Game Theory and Economic Modelling*. Clarendon Press, Oxford, UK.

Lam, S. T. K and Riedl, J. (2009). Is Wikipedia Growing a Longer Tail?. In *Proceedings of the ACM 2009 International Conference on Supporting Group Work (GROUP '09)*. ACM, New York, NY, USA, 105–114. DOI:http://dx.doi.org/10.1145/1531674.1531690

List, C and Pettit, P. (2002). Aggregating sets of judgments: An impossibility result. *Economics and Philosophy* 18, 01 (2002), 89–110.

List, C and Pettit, P. (2004). Aggregating Sets of Judgments: Two Impossibility Results Compared. *Synthese* 140, 1-2 (2004), 207–235.

List, C and Pettit, P. (2011). *Group agency: The possibility, design, and status of corporate agents*. Oxford University Press.

Minsky, M. (1967). *Computation: Finite and Infinite Machines*. Prentice-Hall.

Oizumi, M, Albantakis, L, and Tononi, G. (2014). From the Phenomenology to the Mechanisms of Consciousness: Integrated Information Theory 3.0. *PLoS computational biology* 10, 5 (2014), e1003588.

Reagle, J. M. (2010). *Good Faith Collaboration: The Culture of Wikipedia*. MIT Press.

Sanín, F. G and Wood, E. J. (2014). Ideology in civil war: Instrumental adoption and beyond. *Journal of Peace Research* 51, 2 (2014), 213–226. DOI:http://dx.doi.org/10.1177/0022343313514073

Shalizi, C. R and Moore, C. (2003). What is a macrostate? Subjective observations and objective dynamics. *arXiv preprint cond-mat/0303625* (2003).

Theiner, G. (2008). *From Extended Minds to Group Minds: Rethinking the Boundaries of the Mental*. Ph.D. Dissertation. Indiana University.

Theiner, G. (2013). Onwards and Upwards with the Extended Mind: From Individual to Collective Epistemic Action. In *Developing Scaffolds*, Linnda Caporael, James Griesemer, and William Wimsatt (Eds.). MIT Press, 191–208.

Theiner, G, Allen, C, and Goldstone, R. L. (2010). Recognizing Group Cognition. *Cognitive Systems Research* 11, 4 (2010), 378–395.

Tomasello, M. (2009). *Why We Cooperate*. MIT Press. Responses by Carol Dweck, Joan Silk, Brian Skyrms, and Elizabeth Spelke.

Tononi, G, Edelman, G. M, and Sporns, O. (1998). Complexity and coherency: integrating information in the brain. *Trends in Cognitive Sciences* 2, 12 (1998), 474 – 484. DOI:http://dx.doi.org/10.1016/S1364-6613(98)01259-5

Tononi, G, Sporns, O, and Edelman, G. (1994). A measure for brain complexity: relating functional segregation and integration in the nervous system. *Proceedings of the National Academy of Sciences* 91, 11 (1994), 5033–5037. http://www.pnas.org/content/91/11/5033.abstract

Towne, W. B, Kittur, A, Kinnaird, P, and Herbsleb, J. D. (2013). Your process is showing: controversy management and perceived quality in Wikipedia. In *Computer Supported Cooperative Work, CSCW 2013, San Antonio, TX, USA, February 23-27, 2013*, Amy Bruckman, Scott Counts, Cliff Lampe, and Loren G. Terveen (Eds.). ACM, 1059–1068.

Traulsen, A and Nowak, M. A. (2006). Evolution of cooperation by multilevel selection. *Proceedings of the National Academy of Sciences* 103, 29 (2006), 10952–10955.

Walker, S. I, Cisneros, L, and Davies, P. C. (2012). Evolutionary transitions and top-down causation. *arXiv preprint arXiv:1207.4808* (2012).

Walker, S. I and Davies, P. C. W. (2013). The algorithmic origins of life. *Journal of The Royal Society Interface* 10, 79 (2013). DOI: http://dx.doi.org/10.1098/rsif.2012.0869

Wegner, D. M. (1987). Transactive memory: A contemporary analysis of the group mind. In *Theories of group behavior*. Springer, 185–208.

Welser, H. T, Cosley, D, Kossinets, G, Lin, A, Dokshin, F, Gay, G, and Smith, M. (2011). Finding Social Roles in Wikipedia. In *Proceedings of the 2011 iConference (iConference '11)*. ACM, New York, NY, USA, 122–129. DOI:http://dx.doi.org/10.1145/1940761.1940778

Human Computation (2014) 1:1:31-44
© 2014, Aroyo & Welty. CC-BY-3.0
ISSN: 2330-8001, DOI: 10.15346/hc.v1i1.3

The Three Sides of CrowdTruth

LORA AROYO, VU UNIVERSITY AMSTERDAM

CHRIS WELTY, IBM WATSON RESEARCH CENTER

ABSTRACT

Crowdsourcing is often used to gather annotated data for training and evaluating computational systems that attempt to solve cognitive problems, such as understanding Natural Language sentences. Crowd workers are asked to perform semantic interpretation of sentences to establish a ground truth. The state-of-the-art is for each sentence to be annotated by one worker, and quality is measured as an aggregate property of the entire task, through an average pairwise agreement (κ score) on a small subset of the data that is given to all workers. Our results clearly demonstrate that disagreement indicates low quality, but that quality must be measured on all parts of the annotation task (the workers, the sentences, and the annotation targets). A low aggregate score may indicate bad data, bad task design, bad workers, or some combination. The CrowdTruth approach uses multiple workers on each sentence and gives them multiple annotation choices, which allows us to isolate exactly where problems in an annotation task are, improving the overall quality of the result. We demonstrate this with a set of experiments that show significant improvement in spam detection on workers performing annotation for relation extraction, by including measures of quality for sentences and relations in the measurement of the worker quality.

One popular use of crowdsourcing in AI is to provide a cheaper and more scalable way to gather annotated data for gold standards that are used to train and evaluate machine learning systems. In NLP, crowdsourcing has been used for nearly a decade, as the low level language understanding tasks that are popular research topics often map well into crowdsourcing micro-tasks. However, as we have observed previously (Aroyo and Welty, 2014), the introduction of crowdsourcing hasn't fundamentally changed the way gold standards are created; in particular, humans are asked to provide a semantic interpretation of some data, with an explicit assumption that there is *one correct interpretation*.

We have proposed a new methodology for gathering annotated data from the crowd, inspired by the simple intuition that human interpretation is subjective (Aroyo and Welty, 2013a). From this we have observed that disagreement is a natural product of having multiple people perform annotation tasks, and can provide useful information about the task, a particular annotation unit, or a worker. We propose rejecting the traditional notion of truth in gold standard annotation, in which annotation tasks are viewed as having a single correct answer, adopting instead a disagreement-based approach we call CrowdTruth.

In this paper we explore CrowdTruth in the context of measuring the quality of workers, annotation units, and tasks. We hypothesize that these measures are inter-dependent, and that existing crowdsourcing approaches that measure only worker quality are missing important information, as not all sentences are created equal. We show experimental evidence that these metrics are indeed intertwined, and show improvement by taking that into account. We begin by reviewing human annotation practice for NLP, then we introduce semantic interpretation in general, and the problems that current practices miss. We provide an overiew of CrowdTruth, followed by a survey of the metrics we use, and show experiments that demonstrate the interdependence of quality measurements for workers, annotation units, and target semantics.

Table 1. *Example Sentences and Definitions*

No.	Sentence
ex1	[METHYLERGOMETRINE] is a blood vessel constrictor most commonly used to prevent or control excessive [BLEEDING].
ex2	[GADOLINIUM AGENTS] used for patients with severe renal failure show signs of [NEPHROGENIC SYSTEMIC FIBROSIS].
ex3	He was the first physician to identify the relationship between [HEMOPHILIA] and [HEMOPHILIC ARTHROPATHY].
ex4	[ANTIBIOTICS] are the first line treatment for indications of [TYPHUS].
ex5	Patients with [TYPHUS] who were given [ANTIBIOTICS] exhibited several side-effects.
ex6	With [ANTIBIOTICS] in short supply, DDT was used during World War II to control the insect vectors of [TYPHUS].
ex7	With a [TYPHUS] outbreak, many inhabitants were prescribed [ANTIBIOTICS] without diagnosis.
ex8	[Monica Lewinsky] came here to get away from the chaos in [the nation's capital].

1. BACKGROUND

Machine learning tasks require a gold standard for training, and all cognitivie computing tasks need gold standards for evaluation as well. The simple principle behind human annotation is to have humans perform some semantic interpretation task on data (e.g. audio, video, text, image) to create a reference standard that machines can be compared to.

Consider the NLP task of relation extraction, in which sentences are processed to determine if a particular semantic relation is being expressed in the sentence between two given terms. Table 1 shows a few examples. To create a gold standard, humans are tasked to read each sentence and specify whether a particular semantic relation, such as *TREATS*, is expressed in the sentence between the highlighted terms. Many sentences (hundreds or thousands) are given to these human annotators.

The quality of a human created gold standard is measured in inter-annotator agreement, typically using the κ-coefficient (Cohen, 1960), which is the pairwise disagreement between annotators, corrected for priors, typically reported as the average across all annotators. For the most part, each annotation unit (for relation extraction, this means each sentence) is given to only one person, but some subset of the units are given to everyone in order to calculate the κ score. In different tasks, different ranges of κ scores are considered acceptible; in general a high level of disagreement is considered to be a property of a poorly defined task (Viera and Garrett, 2005).

Often, however, the goal of improving the task definition leads to eliminating disagreement in order to "perfume" the κ-coefficient, hiding the causes for disagreement behind arbitrary decisions that force agreement. This can be seen in most annotation guidelines for NLP tasks, e.g. the MRP Event Extraction Experiment guidelines (Hovy et al., 2012) restrict annotators to follow just one interpretation. For example, spatial information is restricted only to *COUNTRY*, even though other more specific location indicators might be present in the text. The ACE 2002 RDC guidelines V2.3[1] say that "geographic relations are assumed to be static," and claim that sentence *ex8* in Table 1 expresses the *LOCATED* relation between "Monica Lewinsky" and "the nation's capital," even though one clear reading of the sentence is that she is *not* in the capital. Our experiences in designing an annotation task for medical relations had similar results; we found the guidelines becoming more brittle as further examples of annotator disagreement arose. In many cases, experts argued vehemently for certain interpretations being correct, and the decisions made to clarify the "correct" annotation ended up with dissatisfying compromises.

Our work exposed two problems with the ground truth process: the elimination of disagreement was causing the formal task definitions to be overly artificial, and the κ score measure was too course-grained, treating all task components, workers, and annotation units equally.

More recently, as crowdsourcing has increasingly become recognized as useful for creating gold standards, research has focused on measuring the quality of workers, in particular for detecting spam (Alonso and Baeza-Yates, 2011; Raykar and Yu, 2012; Sarasua et al., 2012). This is a productive direction that is not limited to spam detection alone. Although it has not previously been established, it should be clear that *low quality workers generate disagreement*, and whether they are

---

[1] ACE guidelines: http://projects.ldc.upenn.edu/ace/

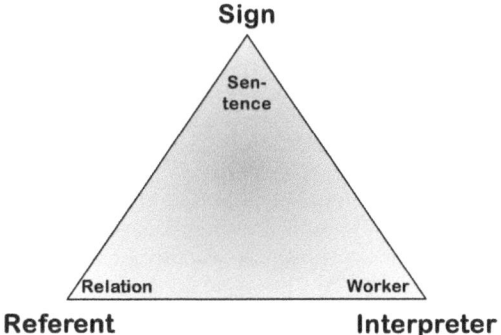

Figure 1. *Triangle of Reference*

intentional spammers, unfamiliar with the task, or simply lazy, they can impact the κ score and make the task appear to be poorly defined.

There are many factors that may cause disagreement, however, and not all of them are indicative of the task. In our work, we sought other aspects of semantic annotation tasks that could impact the quality measurement, and other properties of the annotation results that are useful to measure.

2.   SEMANTIC INTERPRETATION AND ANNOTATION

It is widely accepted that semantic interpretation has three components, illustrated by the *triangle of reference* (see Figure 1) between a sign, something the sign refers to, and the intepreter of the sign (Ogden and Richards, 1923). The interpreter perceives the sign (e.g. a word, a sound, an image, a sentence) and through some cognitive process attempts to find the referent of that sign (e.g. an object, an idea, a class of things). This process of interpretation is what we generally mean when we talk about semantics.

Relation extraction, as discussed above, is an obvious example of semantic interpretation in which sentences such as those shown in Table 1 are the signs, workers are the interpreters, and the referents are provided by the semantics of the domain; in our examples the set of relations are the possible referents, and these are listed in Table 2. The central claim of this paper is that *quality measures should consider all three corners of the triangle of reference*.

An obvious and more widely studied source of quality problems is the interpreters, i.e. the crowd workers or dedicated annotators that provide the semantic intepretation results. However, two less studied parts of the problem come from the other two corners of the triangle.

In Table 1, consider whether examples *ex4-6* express the *TREATS* relation beween the two highlighted terms. In *ex4*, the relation is being expressed directly, and when we give this sentence to multiple annotators, they tend to agree that it expresses the *TREATS* relation. In *ex5*, the *TREATS* relation is not being directly expressed, however a reasonable argument can be made that it is implied. When we give this sentence to multiple annotators, we see some disagreement as to whether *TREATS* is expressed. In *ex6* the sentence requires a deeper justification to interpret the sentence as expressing *TREATS*; one could argue that a shortage of a treatment causes you to eliminate the carriers of a disease. When we give *ex6* to multiple annotators we see even more disagreement than *ex5*. In our experiments we found that *the degree of disagreement can reflect an intuitive ranking of how clearly sentences express a target relation.*

It should be clear that the disagreement between workers in *ex5&6* is not a property of the workers, it is a property of the sentences. They are vague sentences in expressing the *TREATS* relation. Just as a low quality worker can drag down the

Table 2. Relations Set

Abbr	Relation	Definition	Example
sT	TREATS	therapeutic use of an ingredient or a drug	penicillin treats infection
sP	PREVENTS	preventative use of an ingredient or a drug	vitamin C prevents influenza
sD	DIAGNOSE	diagnostic use of an ingredient, test or a drug	RINNE test is used to diagnose hearing loss
sCA	CAUSES	the underlying reason for a symptom or a disease	fever induces dizziness
sL	LOCATION	body part or anatomical structure in which disease or disorder is observed	leukemia is found in the circulatory system
sS	SYMPTOM	deviation from normal function due to disease or abnormality	pain is a symptom of a broken arm
sM	MANIFESTATION	links disorders to the observations that are closely associated with them	liver failure manifests as abdominal distention
sCI	CONTRAINDICATES	a condition that indicates that drug or treatment should not be used	patients with obesity should avoid using danazol
sAW	ASSOCIATED WITH	signs, symptoms or findings that often appear together	patients who smoke often have yellow teeth
sSE	SIDE EFFECT	a secondary condition or symptom that results from a drug or treatment	use of antidepressants causes dryness in the eyes
sIA	IS A	one of the terms is a more specific variation of the other	migraine is a kind of headache
sPO	PART OF	an anatomical or structural sub-component	the left ventricle is part of the heart

task-wide κ score, a bad sentence can do the same. Perhaps more significantly, if an annotator happens to get a batch of vague, confusing, or ambiguous sentences, their own worker score will drop and they may be labelled as spam. We need, therefore, to consider the quality of each sentence when measuring the quality of each worker, and when measuring the quality of the task.

Another potential source of quality problems is the referents themselves. This is the target relation semantics provided to workers in the task definition itself. In Table 1, *ex1* is a sentence that directly expresses the *PREVENTS* relation between the two highlighted terms. When workers are given this sentence and asked if the *PREVENTS* relation is expressed, they tend to agree that it does. When workers are given this sentence and asked if the *TREATS* relation is expressed, some say it does, and some say it does not. Stepping back and considering these two relations, they have two problems: they overlap significantly (many treatments can also prevent a disease) and they are expressed in English similarly; example *ex7* is completely ambiguous with respect to these two relations. When we compare many sentences that workers believe express one of these two relations, we see support for both relations. This is a form of disagreement that reflects on the quality of the relations themselves; workers are being asked to essentially make an arbitrary choice, there is no principled way to distinguish between them in many cases. In our experiments we found that *the degree of disagreement can reflect a semantic or linguistic ambiguity between target relations.*

In general we have found that when the probability of two relations being chosen on the same sentences by a set of workers is high, then the linguistic expression of the relations may be similar, or the relations themselves may be easily confused. If a worker happens to get a batch of sentences that express one of two easily confused relations, their worker score will drop unless it is accounted for.

3. CROWDTRUTH

Our goal is to create a new kind of quality evaluation based on *crowd truth*, in which disagreement is utilized to help understand the annotated instances for training and evaluation (Aroyo and Welty, 2013a). By analogy to image and video tagging games, e.g. Your Paintings Tagger [2] and Yahoo! Video Tag Game (van Zwol et al., 2008), we envision that a crowdsourcing setting could be a good candidate to the problem of insufficient annotation data, however, we do not exploit the typical crowdsourcing agreement between two or more independent taggers, but on the contrary, we harness their disagreement. We allow for a maximum disagreement between the annotators in order to capture a maximum diversity in the relation expressions, based on our hypothesis that disagreement may indicate vagueness or ambiguity in a sentence, in the target semantics being extracted, or may indicate problems with a worker.

---

[2] http://tagger.thepcf.org.uk/

Rel: 15 Workers/sent pair														
Sentence ID	ST	sP	sD	sCA	sL	sS	sM	sCI	sAW	sSE	sIA	sPO	sNONE	sOTH
225527731	0	0	0	1	0	11	0	0	0	0	0	0	0	0
225527732	0	0	0	0	0	7	2	0	2	2	0	1	0	0
225527733	0	0	0	1	0	7	1	0	1	0	0	0	0	1
225527734	0	0	0	0	0	1	0	0	2	0	0	0	0	9
225527735	0	0	0	0	0	13	0	0	0	0	0	0	0	0
225527736	0	0	0	2	0	2	0	0	1	0	0	0	3	4
225527737	0	0	0	2	0	6	2	0	3	1	1	0	0	0
225527738	0	0	0	2	0	0	1	0	0	1	8	1	0	0
225527739	0	0	0	10	0	0	0	0	0	0	0	1	0	0
225527740	0	0	0	10	0	2	1	0	1	0	0	0	0	1
225527741	1	0	0	5	0	3	3	0	1	0	1	0	1	1
225527742	0	0	0	4	0	0	0	0	3	0	0	0	0	4
225527743	0	0	0	1	0	1	2	0	1	0	0	0	0	8
225527744	0	0	0	3	0	1	0	0	1	8	0	0	0	1
225527745	0	0	0	5	0	2	3	0	1	4	0	0	0	0
225527746	0	0	1	1	5	2	0	0	1	0	0	0	2	0
225527747	0	0	0	1	8	2	2	0	1	0	0	0	1	1
225527748	0	0	0	1	7	1	0	0	1	0	0	0	2	1
225527749	0	0	0	0	0	0	0	0	3	0	1	1	4	2
225527750	0	0	0	1	0	4	2	0	3	0	1	2	0	0

Figure 2. Sentence vectors representing crowd annotations on 20 of the 90 sentences, 15 workers per sentence. Rows are individual sentences, columns are the relations. Cells contain the number of workers that selected the relation for the sentence, i.e. 8 workers selected the sIA relation for sentence 738. The cells are heat-mapped per row, highlighting the most popular relation(s) per sentence. Table 2 explains the abbreviations.

We define a crowdsourcing workflow, described in more detail in (Inel et al., 2013). We focused on a set of relations manually selected from UMLS shown in Table 2, with slightly cleaned up glossary definitions of each relation, ignoring relation argument order. The sentences were selected from Wikipedia medical articles using a simple distant-supervision (Mintz et al., 2009) approach that found sentences mentioning both arguments of known relation instances from UMLS. The crowd workers were presented these sentences with the argument words highlighted, and asked to choose all the relations from the set that were expressed in the sentence between the two arguments. They were given two additional options: OTHER, to indicate the argument words were related but not by one of the given relations, and NONE, to indicate that the argument words were not related in the sentence. They were not told the seed relation from UMLS to avoid bias.

Two key points here are: 1) workers are given multiple choices and are allowed to choose any number of them, and 2) that multiple workers are presented the same sentence. This allows us to collect and analyze multiple perspectives and interpretations. To facilitate this, we represent the result of each worker's annotations on a single sentence as a vector of $n + 2$ dimensions, where n is the number of relations + 2 for the NONE and OTHER options. In these vectors, a 1 is given for each relation the worker thought was being expressed, and we use them to form *sentence disagreement vectors* for each sentence by summing all the worker vectors for the sentence. An example set of disagreement vectors are shown in Figure 2.

4. MEASURING CROWDTRUTH

We use the vector representation to measure annotation quality on the three corners of the semantic intepretation triangle: on the workers (for low quality and spam), on the sentences (for clarity), and on the relations (for similarity). Our vector representation for the annotations led us naturally to cosine as a similarity measure, other alternatives are clearly possible, but we have not experimented with them yet. The metrics are discussed in more detail in (Soberón et al., 2013), in this paper we show ways in which these three kinds of metrics are inter-dependent.

4.1. **Worker Metrics**

Worker-sentence disagreement is the average of all the cosine distances between each worker's sentence vector and the full sentence vector (minus that worker). Referring again to Figure 2, a worker who annotated sentences 731 and 732 with sS, which is the most popular choice in each sentence, would have an average cosine distance of $(0.005 + 0.142)/2 = .074$ (low disagreement), whereas a worker who chose sCA for 731 and sM for 732 would have $(1.0 + .870)/2 = .935$ (high disagreement).

Worker-worker disagreement is $1 - avg(\kappa)$ for a particular worker. Since κ is a pairwise metric, we average, for each worker, the κ scores between that worker and all the others. This computation is tricky since each pair of workers will have worked on a different, possibly empty, set of sentences.

The first metric gives us a measure of how much a worker disagrees with the crowd on a sentence basis, and the second gives us an indication as to whether there are consistently like-minded workers. The intuition is that there may be communities of thought that consistently disagree with others, but agree within themselves. Low quality workers generally have high scores in both.

Average relations per sentence is measured for each worker as the number of relations they choose per sentence averaged over all the sentences they annotate. Since the interface allows workers to choose "all relations that apply", a low quality worker can appear to agree more with the crowd by repeatedly choosing multiple relations, thus increasing the chance of overlap. A high score here can help indicate low quality workers.

4.2. **Sentence Metrics**

Sentence-relation score (SRS) is the core crowd truth metric for relation extraction. It is measured for each relation on each sentence as the cosine of the unit vector for the relation with the sentence vector. In Figure 2, the SRS for the sS relation in the first sentence is .996, indicating that relation is very clearly expressed, and .091 for the sCA relation indicating it is not very clearly expressed.

Sentence clarity is defined for each sentence as the max relation score for that sentence. If all the workers selected the same relation for a sentence, the max relation score will be 1, indicating a clear sentence. In Figure 2, sentence 735 has a clarity score of 1, whereas sentence 736 has a clarity score of 0.61, indicating a confusing or ambiguous sentence.

4.3. **Relation Metrics**

Relation similarity (RS) is defined as the *causal power* (Cheng, 1997) $RS(i, j) = [P(R_j|R_i) - P(R_j|\neg R_i)]/[1 - P(R_j|\neg R_i)]$, where $P(R_i)$ is the probability that annotation i appears in a sentence. We want to know if relation R_i is annotated in a sentence, how often relation R_j is as well, but only if R_j is significantly more likely to be annotated when R_i is as well. In Figure 2 we can see that $P(sCA|sS) = .81$, but $P(sCA) = .8$, so this does not indicate a strong association, but $P(sM) = .5$ and $P(sM|sT) = 1.0$, indicating there might be a dependence between them (of course this data set is too small to actually conclude that). A high similarity score indicates the relations are confusable to workers: their semantics may be similar, they may routinely be expressed in similar ways in language, or the semantic specification may be confusing or vague.

Relation ambiguity is defined for each relation as the max relation similarity for the relation. If a relation is clear, then it will have a low score. As noted above, relations that are strongly associated with another may create problems for the annotation task, not to mention for training machines to discern between them.

Relation clarity is defined for each relation as the max sentence-relation score (SRS) for the relation over all sentences. If a relation has a high clarity score, it means that it is at least possible to express the relation clearly. Unclear relations may indicate unattainable NLP targets, problems with the semantic specification, etc. In Figure 2, sS has a relation clarity score of 1.00, which it gets from sentence 735, whereas sT has a relation clarity score of 0.14, from sentence 741.

Relation frequency is the number of times the relation is annotated at least once in a sentence.

5. EXPERIMENTS

We focus on a series of experiments to gather evidence in support of the claim that the three kinds of metrics, representing the three corners of the semantic interpretation triangle, are inter-dependent and influence each other. We show an improvement in quality of spam prediction by considering sentence and relation quality as part of the evaluation of worker quality.

5.1. Data

For the crowd tasks we chose CrowdFlower[3], a cross-platform crowdsourcing service that aggregates smaller regional crowd-sourcing platforms into a single API and toolset. Workers were selected from the U.S, Canada, and the U.K., based only on our desire to have predominatly native English speakers. Over time we maintained a list of worker ids that had been flagged by our system as spammers, and we blocked those ids from further tasks. Workers were paid between $.02 and $.05 per sentence, which varied across the different batches described below. In our extensive tests, including those published here, the price we paid per sentence did not change any of the properties we measured, only the speed at which the task was completed by the crowd (higher paying jobs attract workers more quickly).

Before gathering the judgements on the test set, we performed a series of tests on CrowdFlower to tune several parameters that impact the crowdsourcing results, as described in (Aroyo and Welty, 2013b). Some of the training set data described below came from these experiments.

The data for the main experiments consists of 230 sentences for eight medical seed-relations (treats, prevents, diagnosis, cause, location, symptom, manifestation, disease-has-finding), split into a training set of 140 sentences and a test set of 90, with the seed relations evenly distributed in each set. The seed-relation sentences were generated by a distant supervision technique applied to medical Wikipedia articles; the technique finds sentences in the articles in which related UMLS terms are mentioned. We grouped the sentences into batches of 30, maintaining the even distribution across the seed relations (3-4 per relation per batch). Each batch was run as a seperate job on CrowdFlower. Limiting the batch size allows us to control the impact of spam, since our spam detection metrics currently run offline.

For the test set, we collected 450 judgements (15 per sentence) in each batch (1350 in total), from 63 workers for the first batch, 83 workers for the second and 83 workers for the last. Workers were not allowed to annotate more than 10 sentences in each batch. This measure was imposed in order to make the batches less desireable to spammers, and decrease the bias towards a few participants who annotate all the sentences. A number of workers worked on all three batches, thus the total number of unique workers for all 90 test sentences is 143. From our previous experiences, judgements from workers who annotated two or fewer sentences were uninformative, so we removed these leaving 110 workers and a total of 1292 judgements on the 90 test sentences. The train set was older data, before we established the methodology, it was run as one batch, with 2420 judgements from 429 workers with the same limit of 10 annotations per worker. After removing workers with two or fewer annotations, we had 2222 judgements from 272 unique workers.

In order to support the experiments with spam detection, the crowdsourcing tasks for the 90 sentences were augmented with a justification step for each sentence. Workers had to enter in a text box the words in the sentence that they believed most indicated their selected relations, or an explanation for selecting *NONE* or *OTHER*. The dataset is so small because we manually went through the data and identified low quality workers from their answers; many didn't answer the justification questions, simply copied the entire sentence or random words from the sentence, repeated the same explanations over and over, or their justifications did not make sense. We chose spam detection as a primary experimental basis because the process of creating a ground truth was easiest, although true to the spirit of CrowdTruth there were certainly borderline cases of workers who did not consistently perform the required task, or misunderstood some part of it, but could not clearly be labelled as spam; for these cases we took a majority vote amongst ourselves without looking at the data.

5.2. Worker Quality Baseline

Spam is a part of crowdsourcing and effective elimination of spam must be part of any crowdsourcing platform. As described in our related work section, most research on spam detection assumes micro-tasks have a correct answer – that there is a

---

[3]http://crowdflower.com

Figure 3. *The classification space for spam detection based on a linear combination of the three worker metrics on the 90 sentences. Red squares are low quality (spam) workers, blue diamonds are high quality (not spam). The x axis are the workers, the y axis is the classifier score. The actual score is not relevant, only the relative positions of points in the space with respect to the classification line. Figure (a) shows the space with no filtering of sentences or relations, a single line cannot separate the spammers from non-spammers. Figure (b) shows the space after sentence filtering, Figure (c) after relation filtering, and Figure (d) after both sentence and relation filtering. Sentence filtering makes the classes linearly seperable, and the separation between the classes increases in the subsequent figures.*

ground truth. For CrowdTruth, not only is there no single correct answer but we are interested in the disagreement, which allows us to evaluate all three parts of the semantic interpretation problem.

We examined the worker metrics as a measure of quality (for more details, see (Aroyo and Welty, 2013c)). Our intuition was that low quality workers would disagree consistently with the crowd across all the tasks they performed. A linear combination of the three worker measures trained using 3-fold cross-validation on the train set achieves 93% accuracy on the test set. The classification space is shown in Figure 3a. The space shows five workers who are known spammers (represented as red squares) that were classified as non-spammers, and three workers who were non-spammers (represented as blue dots) that were classified as spammers. The vast majority of non-spammers, however, are quite far from the classification boundary; this indicates our metrics are a strong signal for detecting spam.

5.3. Impact of Sentence Quality on Worker Quality

Our additional intuition is that sentence quality can impact the worker scores. Our initial hypothesis, that disagreement indicates vagueness or ambiguity in sentences, was based on an observation during our attempts to draft annotator guidelines; *the cases where people disagreed were, quite simply, hard to understand*, either because they were vague or ambiguous or difficult to map into the rigid notion of a binary semantic relation. It is reasonable to assume that machines will have just as difficult a time learning from these examples.

Our experiments clearly show that confusing sentences cause disagreement, and it stands to reason that workers who by chance had more than their fair share of confusing sentences will end up looking disagreeable even though they aren't spammers. To test this, we implemented a two-step strategy, first computing the sentence metrics on the sentences and filtering out low quality sentences (one standard deviation below the mean), and then second we re-computed the worker metrics based on the filtered set. In step 1, we filtered out 28 sentences from train and 19 sentences from test.

The resulting classification space is shown in Figure 3b. We do not show the labels for the y-axis in Figure 3, since the re-training generates different weights and the scores are different in the four sub-figures. However, what should be visible is the relative separation of the red and blue dots. Adding the sentence filtering makes the test set 100% classifiable, which is surely an artifact of our small data size, but visible in the figure is the cleaner seperation of the spaces.

Much more importantly, *the worker scores improved in a statistically significant number of cases* ($p < .005$). To determine this, we compared the worker metric scores computed from the full sentence set (baseline), to the scores computed from the filtered set (Sentence Filter). There are two worker metrics calculated from the sentences and discussed in Section 4.1, *Worker-sentence disagreement* and *Worker-worker disagreement*. Each metric is intended to monotonically increase with the liklihood of the worker being low quality, so for each metric we compared each worker's score from the two sentence sets and if the worker was a spammer we subtracted the baseline score from the filtered sentence score, resulting in a positive improvement if the filtered score was higher; if the worker was not a spammer we subtracted the filtered sentence score from the baseline score, resulting in a positive improvement if the filtered score was lower. In Figure 4, the blue lines show the improvement scores for the sentence filtering experiment, note that the baseline is the y-axis. The x-axis represents the workers, but note the three lines shown in each chart are sorted against a different ordering of workers in order to more clearly show the overall differences, thus it is not the case that for every worker the sentence filter improvment is more than the relation filter improvement.

5.4. Impact of Relation Quality on Worker Quality

The central hypothesis of this paper is that all the corners of the semantic interpretation triangle influence each other, and it stands to reason that the referents – the target relations, in this case – will have impact on measurements of the other corners, such as worker quality.

The target relations in our examples, shown in Table 2, were chosen from UMLS for being the most likely to impact our medical question-answering performance in Watson (Ferrucci et al., 2010). Upon consideration, however, it is clear that some of these relations are similar. The relation *CAUSES* is a generalization of the *SYMPTOM* relation, and *MANIFESTATION* is a specialization of it. How are workers to tell them apart in sentences? In our conventional relation annotation task, we tried to dive into the differences and force workers to agree more, for example telling them to use the most specific relation that

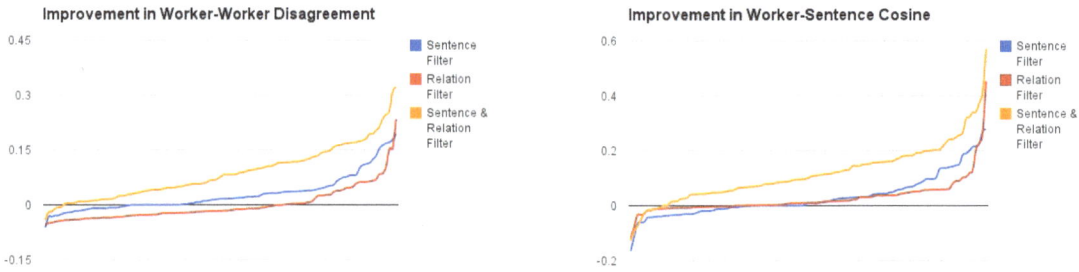

Figure 4. ***Improvements to worker metrics from low-quality sentence filtering, vague and similar relation filtering, and a combination. The x-axis are individual workers, the y-axis is the overall improvement.***

applied. This improved agreement, but still led to confusing cases. We saw no significant difference between the behavior of the crowd and medical experts in this regard (Aroyo and Welty, 2013b).

The most prominent problem in our data is the confusion between *TREATS* and *PREVENTS*. Consider Example *ex1* in Table 1. This sentence generated high disagreement, garnering 8 votes for *TREATS* and 5 for *PREVENTS*. It is difficult to argue for one or the other, indeed in this case the CrowdTruth vector appears to be the only "right" answer. Across our 90 sentence set, 92% of sentences with a *PREVENTS* relation annotated, also had *TREATS*.

Our detailed analysis showed 3 categories of relations that contribute to disagreement:

Vague Relations: Relations with high frequency and low Relation Clarity scores. For example, *ASSOCIATED WITH* appeared in 72% of our sentences but had a Relation Clarity of .61. Workers were simply not able to consistently find examples of it in text. We hypothesize that these relations are just as useful as the *OTHER* category, and that removing them would improve other measures.

Useless Relations: Relations with low frequency and low Relation Clarity basically contribute nothing to the task. In our 90 sentences, *CONTRAINDICATES* appeared in 13% of sentences, and had a relation clarity score of 0.12. It was never the top scoring relation in a sentence. We hypothesize that these relations can be useful for catching spam, as their low coverage in the sentences means that a worker selecting random choices is more likely to choose it than a legitimate worker. However they should be reconsidered in the task design, as they may not be expressible in English. For these experiments we ignored this category and did not test the hypothesis.

Ambiguous Relations: Relations with high Relation Ambiguity scores are easily confused with other relations and may be causing disagreement, such as *TREATS* and *PREVENTS*. We hypothesize that these relations are contributing to disagreement in ways that do not reflect the quality of workers or sentences, and should be merged when computing these other measures.

To measure the impact of relation quality on the worker metrics, we again used a two-step filter and re-compute process, first computing the relation metrics on the original sentence sets, merging the vague relations into *OTHER*, and ambiguous relations with each other. Merging of ambiguous relations was implemented by moving the votes for the lower probability relation with the higher. If a single worker voted for both of a pair of merged relations the second vote was dropped. Merging results in new sentence vectors with fewer dimensions, and in the second step we re-computed the worker metrics based on these new vectors.

In both sets we merged *ASSOCIATED WITH* into *OTHER*, *PREVENTS* into *TREATS*, *SYMPTOM* into *CAUSE*, and *MANIFESTATION* into *CAUSE*.

The results on spam classification are shown in Figure 3c, which should be compared to the spam classification without filters in Figure 3a. As with the use of sentence filters, the relation filtering much more clearly defines the space, with a large separation between positive and negative instances. Again the 100% accuracy is an artifact of the small data size, and the important point is that the pairwise improvements to the worker scores are significant with $p < .006$, slightly less significant that sentence filtering alone. This is shown in Figure 4 as the red line. We see that relation filtering, while having an overall positive effect, has a large number of cases with a very small degradation (negative improvement score) for the worker-worker disagreement metric. The effect of merging relations is to reduce the dimensionality of the comparison space, which overall increases the chances that two workers will agree randomly. However, legitimate worker's behavior is not random, and in the sentences that truly exposed relation similarity many of them would annotate both relations being expressed. For example in sentence *ex7* in Table 1, many workers indicated both *TREATS* and *PREVENTS* as being expressed. In the disagreement score, they are counted as agreeing twice with other workers who behaved the same. When the relations are merged, these workers only get credit for one agreement. So a large number of workers saw very small increases in the disagreement score. Spam workers, who do behave more or less randomly, also tend to benefit slightly from the merging, so they see small decreases in the disagreement score. However, these effects are small. The big winners are workers who were confused by the relation similarity and inconsistently chose one relation or the other in the different sentences they annotated. Thus, the relation filtering improves the false positive rate for spam detection, and the overall improvement is significant.

5.5. Combining Sentence and Relation Filtering

As a final test we combined both the sentence and relation filtering techniques, first filtering out low clarity sentences, then filtering vague and ambiguous relations as described above. The worker metrics were computed on these new sentences and vectors.

The results on spam classification are shown in Figure 3d. The combination proves to even further separate the space. In Figure 4 the orange lines show the dramatic pairwise improvement in worker scores from the baseline (the x-axis). The improvement is significant with $p < .0002$. The improvements over sentence filtering alone or relation filtering alone are both significant ($p < .003$).

6. RELATED WORK

Our work on crowdsourcing as an approach for generating gold standards follows a growing community of machine learning and NLP research (Finin et al., 2010; Chen and Dolan, 2011; Raykar et al., 2010; Snow et al., 2008), e.g. for entity clustering and disambiguation (Lee et al., 2013), as well as for taxonomy creation (Chilton et al., 2013).

The novelty of CrowdTruth is our approach to handling annotator disagreement, that draws some inspiration from existing work. We believe its success is based on a simple observation, that human interpretation is subjective, and that therefore the notion that there can be objective truth for the way people perceive text (or generally for anything outside of mathematics) is flawed. This may seem either obvious or extreme, but the vast majority of research in NLP and data analytics still assumes that for all problems there is a correct answer, and this has led to the current practices of gold standard acquisition. This assumption seems to be so deeply embedded in modern scientific methods that even in cases obviously laden with disagreement, researchers are still building gold standards to train and evaluate machine systems. For example, in (Lee and Hu, 2012), the authors describe an effort to learn mood labels on music, using a large crowdsourced dataset, with the "correct" mood label obtained through majority voting. We suggest that the variety of moods and their popularity would be more useful information than assuming there is a single best mood label in every case.

Similarly, in (Ang et al., 2002) and subsequent work in emotion (Litman, 2004), disagreement is used as a trigger for *consensus-based annotation*. The notion that people can agree on emotion and sentiment outside a very small number of clear cases seems to go against our basic human experience, yet this approach claims very high κ scores (above .9). We suggest it would be extremely useful to additionally examine cases in which people naturally disagreed on emotional labels, and that disagreement would likely indicate interesting borderline cases if not reveal a more accurate model of emotion.

A good survey and set of experiments using disagreement-based semi-supervised learning can be found in (Zhou and Li, 2010), where they use disagreement as a source of varied instance data for bootstrapping. They do not really focus on the disagreement or agreement level, they instead rely on the fact that if the description is relatively unconstrained, people will naturally think of different examples.

We follow a similar strategy for disagreement harnessing in crowdsourcing relation extraction in medical texts as (Chklovski and Mihalcea, 2003) for word sense disambiguation. The authors also form a confusion matrix from the disagreement between annotators, and then use this to form a similarity cluster. Our work adds a classification scheme for annotator disagreement that provides a more meaningful feature space for the confusion matrix, in addition to providing measures of the workers and the relations. Most recently, in (Plank et al., 2014), an approach to dealing with particularly hard examples of part-of-speech tagging is proposed, using an idea similar to our disagreement approach. We believe these efforts add further evidence to our basic hypothesis, that semantic intepretation is subjective, and gathering wide range of human annotations is desireable.

In (Gligorov et al., 2011), their study showed that only 14% of annotations provided by lay users are found in the professional vocabulary (GTAA), which provides a severe limitation of expert-derived vocabularies in supporting user search, since the users don't seem to be able to match the expert terms. Harnessing disagreement brings in multiple perspectives on data, beyond what experts may believe is salient or correct, and may also be of particular value in vocabulary induction, an area of current interest for adapting NLP technology such as Watson to new domains.

When dealing with crowdsourcing, there is a growing literature on observing and analysing workers behaviour (Mason and Suri, 2012) for ultimately being able to detect and eliminate spam (Bozzon et al., 2013; Kittur et al., 2008; Ipeirotis et al., 2010), and analyze workers perfomance for quality control and optimization of the crowdsourcing processes (Singer and Mittal, 2013). Our worker metrics relate to the approach proposed by (Sheng et al., 2008) for improving data quality for supervised learning. Most of the literature on spam detecting again is based on the assumption that for each annotation there is a single correct answer, enabling distance and clustering metrics to detect outliers (Alonso and Baeza-Yates, 2011; Raykar and Yu, 2012; Difallah et al., 2012). We have demonstrated here that spam detection needs to include measures of sentence and relation quality, which has not been previously considered.

CrowdFlower, a popular crowdsourcing platform, implements quality assurance methods based on gold standards, i.e. "golden units" to denote types of questions, for which the answer is trivial or known in advance. For example, CROWDMAP (Sarasua et al., 2012) uses golden units to block invalid answers, as well as use verification questions that force the user to type a name of the selected concept. This is effective in eliminating automated workers. Additionally CrowdFlower allows for filtering spammers at run time based on country or previously built trust calculating mechanisms(Oleson et al., 2011).

In the case of crowdsourcing ground truth data, the correct answer is not known, thus building golden units is difficult. As (Bachrach et al., 2012) points out, a possible solution from phycology research could be to evaluate responses to items for which the correct answer is known (Anastasi and Urbina, 1997), or alternatively, as proposed by the DARE model in (Bachrach et al., 2012), to use graphical models to infer the correct answer for each question (when these are not known in advance).

In all these cases, whether known or not, the assumption that there is a correct answer for each micro-task is explicit. However, as discussed above, our claim goes further, based on experimental results showing that often *there is not only one correct answer*, which changes the kind of modeling required to detect spam. We have shown some effective metrics and continue to explore the possibilities.

In previous work on similar measures for folksonomies (Markines et al., 2009), the authors review and evaluate a suite of similarity measures for users, resources, and tags in the social tagging setting. While we evaluate our framework only for creating a relation extraction gold standard, the approaches share a great deal of similarity. To begin with, the idea that three factors contribute to the interpretation: the workers, the thing being interpreted (the resource or the sentence), and the domain semantics (the set of tags or the set of relations), which is reminiscent of the semiotic triangle (Ogden and Richards, 1923). In both cases the rich diversity of the crowd's input is viewed as desirable, though this is less of a departure from the state of practice in the folksonomy community than in NLP. We believe both approaches generalize into a richer framework, in which similarity and disagreement are complementary tools for gaining further insight into the semantics that can be gleaned from crowds.

7. CONCLUSIONS

We have previously proposed a new approach to human annotation of gold standard data for relation extraction components, that we believe generalizes to problems for which a gold standard is needed for training and evaluation. Our CrowdTruth approach promises to be faster, cheaper, and more scalable that traditional ground truth approaches involving dedicated human annotators, by exploiting the disagreement between crowd workers as a signal, rather than trying to eliminate it. The basis of our approach is to have multiple workers annotate the same sentence, and allow them multiple annotation choices for each one. This gives us useful data to measure quality of the resulting annotated data.

In previous work we showed that the quality of CrowdTruth is comparable to expert human annotators (Aroyo and Welty, 2013b), and that disagreement can be a useful signal in detecting low quality workers (Soberón et al., 2013). In this paper we have shown evidence that quality measures in semantic interpretation tasks are inter-dependent, and higher accuracy can be achieved by considering the impact of sentence quality and relation quality on worker quality measurements. We showed significant improvement in worker quality metrics with respect to known spammers by incorporating the quality of the individual sentences and target relations.

This is drastically different than the state-of-the-art in human annotation, where each sentence is annotated by one worker, and quality is measured as an aggregate property of the entire task, through an average pairwise agreement (κ score) on a small subset of the data that is given to all workers. Our results clearly demonstrate that disagreement indicates low quality, but that quality must be measured on all parts of the annotation task (the workers, the sentences, and the annotation targets). A low aggregate score may indicate bad data, bad task design, bad workers, or some combination. The CrowdTruth approach is able to isolate exactly where problems in an annotation task are, improving overall quality of the result.

In future work we plan to more exhaustively explore the relationships between the different corners of the *triangle of reference*, e.g. the impact of relation and worker quality on sentence measures, and of worker and sentence quality on relation measures. We are also working on making our framework open-source, and available as a service, in order to generalize CrowdTruth to other domains. For more information about CrowdTruth, see http://crowdtruth.org.

8. REFERENCES

Alonso, O and Baeza-Yates, R. (2011). Design and implementation of relevance assessments using crowdsourcing. In *In Proc. ECAIR*. Springer-Verlag, 153–164. http://dl.acm.org/citation.cfm?id=1996889.1996910

Anastasi, A and Urbina, S. (1997). *Psychological testing*. Prentice Hall. http://books.google.nl/books?id=lfFGAAAAMAAJ

Ang, J, Dhillon, R, Krupski, A, Shriberg, E, and Stolcke, A. (2002). Prosody-Based Automatic Detection Of Annoyance And Frustration In Human-Computer Dialog. In *in Proc. ICSLP 2002*. 2037–2040.

Aroyo, L and Welty, C. (2013)a. Crowd Truth: Harnessing disagreement in crowdsourcing a relation extraction gold standard. In *Web Science 2013*. ACM.

Aroyo, L and Welty, C. (2013)b. *Harnessing Disagreement in Crowdsourcing a Relation Extraction Gold Standard*. Technical Report No.203386. IBM Research.

Aroyo, L and Welty, C. (2013)c. Measuring Crowd Truth for Medical Relation Extraction. In *AAAI 2013 Fall Symposium on Semantic for Big Data*. AAAI.

Aroyo, L and Welty, C. (2014). Truth is a Lie: Seven myths about human annotation. *AI Magazine* (2014).

Bachrach, Y, Graepel, T, Minka, T, and Guiver, J. (2012). How To Grade a Test Without Knowing the Answers - A Bayesian Graphical Model for Adaptive Crowdsourcing and Aptitude Testing.. In *ICML*. icml.cc / Omnipress.

Bozzon, A, Brambilla, M, Ceri, S, and Mauri, A. (2013). Reactive crowdsourcing. In *Proceedings of the 22nd international conference on World Wide Web (WWW '13)*. International World Wide Web Conferences Steering Committee, Republic and Canton of Geneva, Switzerland, 153–164. http://dl.acm.org/citation.cfm?id=2488388.2488403

Chen, D and Dolan, W. (2011). Building a Persistent Workforce on Mechanical Turk for Multilingual Data Collection. (2011). http://citeseerx.ist.psu.edu/viewdoc/download?rep=rep1&type=pdf&doi=10.1.1.222.595

Cheng, P. (1997). From covariation to causation: A causal power theory. *Psychological Review* 104 (1997), 367âĂŞ405.

Chilton, L. B, Little, G, Edge, D, Weld, D. S, and Landay, J. A. (2013). Cascade: crowdsourcing taxonomy creation. In *Proceedings of the SIGCHI Conference on Human Factors in Computing Systems (CHI '13)*. ACM, New York, NY, USA, 1999–2008. DOI:http://dx.doi.org/10.1145/2470654.2466265

Chklovski, T and Mihalcea, R. (2003). Exploiting Agreement and Disagreement of Human Annotators for Word Sense Disambiguation. In *UNT Scholarly Works*. UNT Digital Library. http://digital.library.unt.edu/ark:/67531/metadc30948/

Cohen, J. (1960). A coefficient of agreement for nominal scales. *Educational and Psychological Measurement* 20 (1960), 37–46.

Difallah, D. E, Demartini, G, and Cudré-Mauroux, P. (2012). Mechanical Cheat: Spamming Schemes and Adversarial Techniques on Crowdsourcing Platforms. In *CrowdSearch*. 26–30.

Ferrucci, D, Brown, E, Chu-Carroll, J, Fan, J, Gondek, D, Kalyanpur, A. A, Lally, A, Murdock, J. W, Nyberg, E, Prager, J, Schlaefer, N, and Welty, C. (2010). Building Watson: An Overview of the DeepQA Project. *AI Magazine* 31 (2010), 59–79. Issue 3.

Finin, T, Murnane, W, Karandikar, A, Keller, N, Martineau, J, and Dredze, M. (2010). Annotating named entities in Twitter data with crowdsourcing. In *In Proc. NAACL HLT (CSLDAMT '10)*. Association for Computational Linguistics, 80–88.

Gligorov, R, Hildebrand, M, van Ossenbruggen, J, Schreiber, G, and Aroyo, L. (2011). On the role of user-generated metadata in audio visual collections. In *K-CAP*. 145–152.

Hovy, E, Mitamura, T, and Verdejo, F. (2012). *Event Coreference Annotation Manual*. Technical Report. Information Sciences Institute (ISI).

Inel, O, Aroyo, L, Welty, C, and Sips, R.-J. (2013). Exploiting Crowdsourcing Disagreement with Various Domain-Independent Quality Measures. In *Proceedings of the 3rd International Workshop on Detection, Representation, and Exploitation of Events in the Semantic Web (DeRiVE 2013), 12th International Semantic Web Conference*.

Ipeirotis, P. G, Provost, F, and Wang, J. (2010). Quality management on Amazon Mechanical Turk. In *Proceedings of the ACM SIGKDD Workshop on Human Computation (HCOMP '10)*. ACM, New York, NY, USA, 64–67. DOI:http://dx.doi.org/10.1145/1837885.1837906

Kittur, A, Chi, E. H, and Suh, B. (2008). Crowdsourcing user studies with Mechanical Turk. In *Proceedings of the SIGCHI Conference on Human Factors in Computing Systems (CHI '08)*. ACM, New York, NY, USA, 453–456. http://doi.acm.org/10.1145/1357054.1357127

Lee, J, Cho, H, Park, J.-W, Cha, Y.-r, Hwang, S.-w, Nie, Z, and Wen, J.-R. (2013). Hybrid entity clustering using crowds and data. *The VLDB Journal* 22, 5 (2013), 711–726. DOI:http://dx.doi.org/10.1007/s00778-013-0328-8

Lee, J. H and Hu, X. (2012). Generating ground truth for music mood classification using mechanical turk. In *Proceedings of the 12th ACM/IEEE-CS joint conference on Digital Libraries (JCDL '12)*. ACM, New York, NY, USA, 129–138. DOI:http://dx.doi.org/10.1145/2232817.2232842

Litman, D. J. (2004). Annotating Student Emotional States in Spoken Tutoring Dialogues. In *In Proc. 5th SIGdial Workshop on Discourse and Dialogue*. 144–153.

Markines, B, Cattuto, C, Menczer, F, Benz, D, Hotho, A, and Stumme, G. (2009). Evaluating similarity measures for emergent semantics of social tagging. In *Proceedings of the 18th international conference on World wide web (WWW '09)*. ACM, New York, NY, USA, 641–650. DOI:http://dx.doi.org/10.1145/1526709.1526796

Mason, W and Suri, S. (2012). Conducting behavioral research on AmazonâĂŽĂĎÃt's Mechanical Turk. *Behavior Research Methods* 44, 1 (2012), 1–23. DOI: http://dx.doi.org/10.3758/s13428-011-0124-6

Mintz, M, Bills, S, Snow, R, and Jurafsky, D. (2009). Distant supervision for relation extraction without labeled data. In *In Proc. ACL and Natural Language Processing of the AFNLP: Vol2*. Association for Computational Linguistics, 1003–1011.

Ogden, C. K and Richards, I. (1923). *The meaning of meaning*. Trubner & Co, London.

Oleson, D, Sorokin, A, Laughlin, G. P, Hester, V, Le, J, and Biewald, L. (2011). Programmatic Gold: Targeted and Scalable Quality Assurance in Crowdsourcing. In *Human Computation*.

Plank, B, Hovy, D, and SÃ¥ygaard, A. (2014). Learning part-of-speech taggers with inter-annotator agreement loss. In *Proceedings of EACL-2014*.

Raykar, V. C and Yu, S. (2012). Eliminating Spammers and Ranking Annotators for Crowdsourced Labeling Tasks. *J. Mach. Learn. Res.* 13 (March 2012), 491–518. http://dl.acm.org/citation.cfm?id=2188385.2188401

Raykar, V. C, Yu, S, Zhao, L. H, Valadez, G. H, Florin, C, Bogoni, L, and Moy, L. (2010). Learning From Crowds. *Journal of Machine Learning Research* 11 (2010), 1297–1322.

Sarasua, C, Simperl, E, and Noy, N. F. (2012). CrowdMap: Crowdsourcing Ontology Alignment with Microtasks. In *International Semantic Web Conference (1)*. 525–541.

Sheng, V. S, Provost, F, and Ipeirotis, P. G. (2008). Get another label? improving data quality and data mining using multiple, noisy labelers. In *Proceedings of the 14th ACM SIGKDD international conference on Knowledge discovery and data mining (KDD '08)*. ACM, New York, NY, USA, 614–622. DOI:http://dx.doi.org/10.1145/1401890.1401965

Singer, Y and Mittal, M. (2013). Pricing mechanisms for crowdsourcing markets. In *Proceedings of the 22nd international conference on World Wide Web (WWW '13)*. International World Wide Web Conferences Steering Committee, Republic and Canton of Geneva, Switzerland, 1157–1166. http://dl.acm.org/citation.cfm?id=2488388.2488489

Snow, R, O'Connor, B, Jurafsky, D, and Ng, A. Y. (2008). Cheap and fast—but is it good?: evaluating non-expert annotations for natural language tasks. In *Proceedings of the Conference on Empirical Methods in Natural Language Processing (EMNLP '08)*. Association for Computational Linguistics, Stroudsburg, PA, USA, 254–263. http://dl.acm.org/citation.cfm?id=1613715.1613751

Soberón, G, Aroyo, L, Welty, C, Inel, O, Lin, H, and Overmeen, M. (2013). Measuring Crowd Truth: Disagreement Metrics Combined with Worker Behavior Filters. In *Proceedings of the 1st International Workshop on Crowdsourcing the Semantic Web (CrowdSem 2013), 12th International Semantic Web Conference*.

van Zwol, R, Garcia, L, Ramirez, G, Sigurbjornsson, B, and Labad, M. (2008). Video Tag Game. In *WWW Conference, developer track)*. ACM.

Viera, A. J and Garrett, J. M. (2005). Understanding interobserver agreement: the kappa statistic. *Family Medicine* 37, 5 (2005), 360–363.

Zhou, Z.-H and Li, M. (2010). Semi-supervised learning by disagreement. *Knowl. Inf. Syst.* 24, 3 (2010), 415–439.

Human Computation (2014) 1:1:45-65
ISSN: 2330-8001, DOI: 10.15346/hc.v1i1.4

Toward Complexity Measures for Systems Involving Human Computation

R. JORDAN CROUSER, TUFTS UNIVERSITY

BENJAMIN HESCOTT, TUFTS UNIVERSITY

REMCO CHANG, TUFTS UNIVERSITY

ABSTRACT

This paper introduces the Human Oracle Model as a method for characterizing and quantifying the use of human processing power as part of an algorithmic process. The utility of this model is demonstrated through a comparative algorithmic analysis of several well-known human computation systems, as well as the definition of a preliminary characterization of the space of human computation under this model. Through this research, we hope to gain insight about the challenges unique to human computation and direct the search for efficient human computation algorithms.

1. INTRODUCTION

Computational complexity theory is a branch of theoretical computer science dedicated to describing and classifying computational problems according to their fundamental difficulty, which we measure in terms of the resources required to solve them. One way to measure a problem's difficulty is with respect to time; we may ask **how many operations do I need to perform to find an answer**? Alternatively, one might want to measure difficulty in terms of space; here we could ask **how much memory will I need to execute this process**? These questions, which do not rely on the specific implementation details, are at the heart of computer science. Theoretical arguments ground our intuitions about the problem space, and pave the way for us to design future systems that make these provably correct solutions tractable.

To date, the field of human computation has concerned itself almost entirely with questions of *tractability*. That is, can using human computation make it possible to efficiently solve problems whose solutions are otherwise thought to be too expensive to compute? Using experiential knowledge regarding the kinds of processing that humans are "better" at, such as recognizing objects and speaking naturally, we build systems that capitalize on these skills and offer them as constructive proof: tangible evidence that the problems are in fact tractable using human computation, even when other methods have failed. As such, the development of real-world implementations has far outpaced the development of theoretical measures. Many of these implementations have demonstrated unparalleled success at problems previously thought to be intractable, such as protein folding (Cooper et al., 2010). However, in the absence of a rigorous theory in which to ground new algorithms, researchers must rely on intuition and some deeply-rooted assumptions about the differences between human and machine computation. Extending complexity models to encompass both human and machine computation will help us understand our successes and failures on existing problems, and more fully illuminate the computational space they inhabit.

Computational complexity theory takes the study of solvable problems to a deeper level by asking about the **resources** needed to solve them in terms of time and memory. It enables us to ask questions that get at the fundamental nature of the problem and how we might go about solving it more effectively. Does randomization help? Can the process be sped up using parallelism? Are approximations easier? By understanding the resources required, we can begin to group problems into *complexity classes*, with members of the same class requiring similar kinds or quantities of resources. It also enables us to investigate the effect of limiting these resources on the classes of tasks that can still be solved.

Until now, no such theoretical apparatus has existed for human computation. This is due in large part to the fact that our ability to model **how** the human brain computes is hindered by a limited understanding of the biological mechanisms that enable that computation. While cognitive modeling techniques can help us to understand the interplay between stimulus and response, existing modeling tools are not designed to emulate the *complexity* of the model itself. Until our understanding of the cognitive processes involved in computation is more fully developed, it seems likely that human computation will generally remain a (somewhat finicky) black box in the larger system diagram. In the interim, we can begin to to characterize and quantify the use of human processing power as part of an algorithmic process, rather than waiting to model and measure the cost of the humans' computational processes themselves. By separating questions of per-operation cost from questions of resource utilization, we posit that such models will be robust even as more nuanced and complete models of the human brain come to light.

In this work, we introduce the notion of Human computational complexity theory - an extension of traditional computational complexity theory that accounts for (some of) the vagaries human participatory behavior. We present the Human Oracle Model as a novel method for characterizing and quantifying the use of human processing power as part of an algorithmic process. We then use this model to compare and analyze several well-known human computation systems in the area of Image Labeling, which serves as a vignette to demonstrate the kinds of low-level algorithmic comparisons which might be possible under an abstract model. Finally, we describe how this model can be used to characterize the space of human computation, as well as discuss the model's limitations and its potential for broader impact.

We argue that in order for human computation to achieve legitimacy as a breakthrough computational paradigm, it is imperative that we as a community establish mechanisms by which we abstract the underlying processes away from the details of the implementation and *reason* about the computation being done in human computation systems. The examples presented herein are just a small scratch on the surface of that much larger problem. Their purpose is not to exhaustively illuminate the landscape of human computation problems, but to illustrate that such reasoning is possible using only minor modifications to tools with which computational scientists are already familiar. Through this research, we hope to gain insight about issues unique to this problem space and direct the search for more efficient human computation solutions. In addition, we hope to set the stage for continued dialogue between the human and traditional computation communities, that together we may move forward in pushing the bounds of what is possible and challenge our deep-seated assumptions about what it means to compute.

2. FOUNDATIONS AND RELATED WORK

In traditional computer science, the field of **computational complexity** compares computational processes by evaluating the time and space required to solve a well-specified problem using a computer (for a detailed introduction to computational complexity, please see (Arora and Barak, 2009)). In many cases, it is particularly useful to consider the boundary cases of such problems – that is, what is the best we can hope to do under ideal conditions, and what kind of performance can be guaranteed in the worst case? By better understanding these boundary cases, we become able to characterize the practical limits of what computational processes can do, as well as better isolate the underlying subtasks that make some computational problems truly difficult to solve. When theoretical modeling is applied to systems that involve humans, we posit that it is similarly useful to begin by establishing bounds on clearly-defined edge cases, and we can remain agnostic to behaviors for which boundaries are currently ill-defined so long as we understand how such behavioral processes affect overall system behavior in the limit.

Under existing computational models, many interesting real-world problems are known to be computationally infeasible, even if the path to finding the solution is clear. For example, we know how to solve the Traveling Salesman problem, but computing the solution is intractable for all but a handful of special cases (Lawler et al., 1985). Other problems, such as general image recognition, have no known computational solution at all. In contrast, humans appear to perform remarkably well on many such problems with relative ease (MacGregor and Ormerod, 1996). Robust perceptual systems and learned experiences offer humans a distinct advantage in areas such as visual classification, a fact which lies at the heart of the emerging field of human computation. While our understanding of the biological mechanisms that enable computation in the human brain is still limited, there is an intuition that human computational processes are different from, and in may cases complementary to, mechanical computation.

One recent framework (Crouser and Chang, 2012) illustrates this complementarity, and attempts to organize existing literature on human-machine collaborative systems according to which skills, or affordances, the system leverages. Other taxonomies (Bertini and Lalanne, 2010; Quinn and Bederson, 2011) propose additional classification dimensions such as *human-machine balance*, *motivation*, *aggregation*, *quality control*, *process order*, and *task-request cardinality*. While these frameworks provide a vocabulary for describing human-computer collaborative systems, they fall short of enabling us to quantify the computational work being done in the underlying human computation algorithms. Consider a sample of the numerous published human computation systems in the area of Image Labeling: the ESP Game (von Ahn and Dabbish, 2004), KissKissBan (Ho et al., 2009) and Peekaboom (von Ahn et al., 2006). Each employs human visual perception and linguistic ability to process and describe images, and uses the machine to distribute tasks and aggregate the results. Each uses entertainment as a primary motivator, and redundancy to ensure validity of the resulting data. Given the similarity of the problem as well as the approach to solving it, how do the underlying algorithms compare? To date, there exists no mechanism for performing such a comparison.

Establishing bounds on algorithmic processes and deepening our understanding of the relationships among the problems they solve are of critical importance to the study and design of systems involving human computation. Drawing parallels at the algorithmic level rather than at the implementation level enables us to compare solutions more rigorously than using simple A-B testing. As with other branches of computational science, identifying areas where existing algorithms are redundant or inefficient will enable the design of more efficient algorithms in the future. In addition, reporting bounds on the complexity of human computation algorithms along with the observed performance of the system would improve study reproducibility, as well as help isolate the effects of interface design and other implementation details.

In order to facilitate these kinds of comparisons, it is important to extend traditional theoretical abstractions of computation to capture some of the characteristic differences between computing on silicon and computing using human brainpower. This is a daunting challenge: precisely which facets of "humanness" are the critical components which must be captured in order for a computational model to be both robust and useful? We intentionally err on the side of simplicity in this early attempt, looking to existing work in measuring intelligent contribution to a computational process from the field of Artificial Intelligence as a foundation and asking what we might learn about human computation from even the simplest of theoretical abstractions. While we most certainly fall short of solving the problem of measuring the complexity in human computational systems, it is our hope that this preliminary work will provoke both thought and debate about the role of computational complexity in human computation, and vice versa.

2.1. **Computation with Human Oracles**

Research in the field of Artificial Intelligence seeks to model and emulate human intelligence using a machine. Research in human computation leverages *actual* human intelligence to perform computationally-difficult tasks. Both fields hinge on the long-held belief that there exist problems that require human-level intelligence and reasoning to solve. Because of this relationship, we believe that theoretical models from the Artificial Intelligence community may be a useful starting point for understanding and comparing human computation problems and their solutions.

Theoretical computer science uses abstract models of computational systems, such as Turing Machines (Turing, 1938), to simulate computational processes[1] and explore the limits of what can be computed. In some cases, it is useful to allow the Turing Machine to query an Oracle – a superpowerful subroutine which is able to decide specific problems with perfect accuracy in constant time. Under this model, the overall cost of running the algorithm is the machine's computation plus the time required to generate queries to the Oracle; the Oracle's computation is performed for free. While the computational machinery used by the Oracle to perform its computation is a black box, we can characterize the amount of work the Oracle is being asked to perform by counting the

---

[1] We have elected to begin by expanding on the standard notion of the Turing Machine in this early work due to its elegance, completeness, and relative ubiquity in the discussion of classical computational complexity. It has been suggested that other models, such as non-deterministic Turing Machines, Markov Chain models or Abstract State Machines more generally, might also be useful in exploring the edges of human computation. We hope that the utility of these and other models will be explored in future work.

number of queries the Oracle must answer in order for the computation to terminate. This is known as *query complexity*.

Shahaf and Amir proposed an extension to the standard computational model in which questions may be asked of a Human Oracle – an Oracle with human-level intelligence (Shahaf and Amir, 2007). In this model, the Human Oracle is able to answer questions to which a human would be able to respond, even if a machine could not. They suggest that the complexity of an algorithm executed on such a machine can be represented as a pair $\langle \Phi_H, \Phi_M \rangle$, where Φ_H indicates the query complexity of the algorithm (number of queries to the Human Oracle) as a function of the input size, and Φ_M is the the complexity of the computation performed by the machine. Whenever the complexity of the machine's computation is the same, the complexity of two algorithms can be compared by considering which has a higher query complexity. The minimal complexity of a problem can then be thought of as the minimization of both human and machine cost over all algorithms that correctly solve the problem.

2.2. Value of an Oracle Model for Human Computation

Modeling the human contributions to an algorithm as queries to an Oracle captures the underlying behavior of many existing human computation algorithms. For example, in the well-studied ESP Game (von Ahn and Dabbish, 2004) a human is given some input (an image) and, like an Oracle, is expected to provide a (relatively) correct response to exactly one question: *What do you see?* This interchange, where an external entity is used to inexpensively perform some challenging subroutine, is exactly the kind of system that Oracle machines were designed to describe. Because of this, we adopt the Human Oracle Model as a preliminary mechanism to make quantitative comparisons among human computation algorithms.

Despite the simplicity of the Human Oracle Model, this level of abstraction has several benefits. First, it enables a direct quantification of the cost of an algorithm leveraging human-level intelligence, or human computation, in terms of the number of queries made to the human. This enables a straightforward comparison between two human computation solutions to a given problem on a given input. Second, it enables an objective theoretical comparison between algorithms using humans and the best known purely mechanical algorithms, if they exist. Finally, it separates implementation-specific details such as error control, motivation, and interface design from the algorithm itself. This is an important differentiation, and much in keeping with the spirit of traditional complexity models wherein the performance of an algorithm is assessed independent of the languages in which it may later be implemented or the hardware on which it may be run. While questions of recruiting and incentivizing human contributors is by no means unimportant, we specifically investigate the complexity of the underlying algorithms independently.

Technically speaking, a human can simulate any process the machine can execute. Given an understanding of the process, enough paper and a sufficient supply of pencils, a human operator could write out the contents of each register, perform each bitwise operation, and record each result by hand. However, the time and resources required by the human to compute exactly the same result would be exorbitant. In addition, humans are susceptible to fatigue, and are arguably limited by unreliable recall and the capacity of working memory. In this sense, human operations are expensive.

Because of this, there is an implicit assumption that the use of human processing power in such systems will be judicious. After all, there is a point at which human "processors" will simply refuse to perform any more computation. Much effort has been put into learning how to best incentivize human processors to perform computation (Mason and Watts, 2010; Singer and Mittal, 2011). *Games with a Purpose* try to make participation more entertaining for the human (von Ahn, 2006), thereby increasing their willingness to contribute. However, to date few mechanisms have been developed for comparing the algorithmic processes underlying human computation systems independent of the details of their implementation. While questions of recruiting and incentivizing human contributors are by no means unimportant, in this work we specifically investigate the complexity of the underlying processes at the algorithmic level independent of these factors.

3.   ADAPTING THE HUMAN ORACLE MODEL FOR HUMAN COMPUTATION

Throughout the remainder of this paper, we will adopt two slight relaxations of the assumptions about the behavior of the Human Oracle as differentiated from traditional set-theoretic definitions of an Oracle. These relaxations are meant to enable meaningful comparison between simple *Games with a Purpose*-style human computation system by capturing a small part of the variability inherent in leveraging humans in these computational processes, such as fuzzy interpretations of "correct" responses and the existence of short term memory. While we do not suggest that the following model is complete, we hope that it will provide a toehold for future exploration in this important area.

3.1.   Variability in Human Oracle Responses

Under the standard set-theoretic definition of an Oracle, any two Oracles to the same problem are equivalent with respect to the answers they return to a given query. In contrast, one could reasonably expect that different Human Oracles to the same problem may return different answers when queried on the same input, if more than one appropriate answer exists. Whenever there is potential for ambiguity in processing stimuli, there may be more than one valid response for any given input. However, a given individual may strongly favor one response over another.

We characterize this behavior as follows. Under this model, we will assume that there exist **finitely many** reasonable responses for any query/input pairing:

$$R_Q(x) = \{r_1, r_2, \ldots, r_{n-1}, r_n\}$$

where r_i is a reasonable response to query Q on input x. We then state that any valid Human Oracle always returns one such reasonable answer, but that we can't predict which one they may decide to return. We can express this behavior by defining the Human Oracle H as having a probability distribution over the collection $R_Q(x)$:

$$D_{H(Q,x)} = \{\langle r, P_H(r_i)\rangle | r_i \in R_Q(x), 0 \le r_i \le 1\}$$

where $P_H(r_i)$ is the probability that Human Oracle H returns response r_i when passed query Q on input x, and $\sum_{i=1}^{n} P_H(r_i) = 1$.

In the simplest case, $n = P_H(r_n) = 1$. That is, if there is only one reasonable response, the Human Oracle will return that response with probability 1. When there are multiple reasonable responses, the Human Oracle's probability distribution may heavily favor some subset of responses. We suggest that this probabilistic behavior helps capture the influence of individual differences inherent in any human population. These inconsistencies may be due to different lived experiences, internal biases, or preferences. In addition to individual differences, this distribution may be influenced through incentivization. This may happen *a priori*, such as in systems that incentivize the generation of short responses over more verbose ones, or the distribution may be changed *on-the-fly*, such as in gameified systems where the players may be asked to match (or avoid matching) a partner's responses.

For simplicity, we will exclude nonterminating algorithmic executions from consideration under this model by assuming that the intersection between any two Human Oracle's response sequences will be nonempty, provided they are queried sufficiently many times on the same input. This helps capture the notion of *collective intelligence* relied upon in many human computation applications, and enables meaningful comparisons between the worst-case performance of various approaches. When such an assumption is not made, the worst case performance of any algorithm that utilizes output agreement is trivially infinite; by repeatedly matching pairs of Human Oracles with non-overlapping response sets, the system could continue to query without ever being able to validate an answer.

In practice, individual differences sometimes dictate that the probability of intersection between two people's response sequences is arbitrarily small. For example, one partner may be unfamiliar with an animal not native to their region or a landmark they have never seen, and may therefore not describe it using the same terminology as their more familiar collaborator. To deal with this reality, many real-world systems implement a timeout mechanism to ensure that an algorithm does not spend too much time on a potentially fruitless pairing of Human Oracles. While beyond the scope of this paper, it is also interesting to consider the effects of relaxing the nonempty intersection assumption on algorithmic analysis and expected performance when given some additional information about the distribution of knowledge in the sample population.

3.2. Persistence of Previous Responses

If the same Human Oracle is queried more than once on the same input during the execution of an algorithm, we may wish to assume that it will be aware of its previous responses and will only return each answer once. This is akin to assuming that upon reading the input, the Human Oracle constructs a predefined sequence of answers by ordering their possible responses in decreasing order of probability:

$$A_{H(Q,x)} = (a_1, a_2, \ldots, a_n)$$

where:

$$P(a_{i+1}) < P(a_i) \; \forall \; 1 \leq i \leq n$$

The Human Oracle will answer each query about that particular input by simply reporting the next unused element in the sequence. This reflects human short-term memory, and can be simulated by recording previous responses in the main algorithm and passing the entire history back as part of the input to a non-persistent Oracle.

3.3. **Additional Assumptions**

Additionally, we presume that the Human Oracle can efficiently generate an answer to the queries we pose. In traditional computational models, it is assumed that the Oracle can interpret and respond correctly to the query in constant time. However, it is also acceptable to consider Oracles with other (bounded) response time complexities. With Human Oracles, we do not necessarily know how long it takes a person to solve the problem. For simplicity, we will assume a constant cost for each query to the Human Oracle given a particular affordance, which enables us to consider the complexity of two algorithms leveraging the same kind of Human Oracle in terms of the number of queries required.

Finally, the study of human computation presumes the existence of problems for which humans are *faster, more accurate, more robust, or otherwise superior to* any known machine algorithm. To that end, we only consider problems in which the Human Oracle's answers are integral to computing the solution. That is, the algorithm querying the Human Oracle cannot efficiently generate answers to its own queries, and must rely on (and potentially validate) the responses it receives from the Human Oracle.

We believe that these adaptations result in a model that more closely resembles observed behavior in systems involving human computation, and help capture some of the ambiguity inherent in many interesting human computation problems. In the following section, we use this model as a lens to explore various problems that fall under the umbrella of image labeling. We do not mean to imply that these are necessarily "canonical" or "complete" problems for human computation, as this concept is yet ill-defined. However, we believe that a close examination of well-studied problems through this lens may provide insight into the structure of human computation algorithms. We hope that this will serve as an initial benchmark by which other problems may be measured as we continue to explore the space of human computation.

4. **IMAGE LABELING UNDER THE HUMAN ORACLE MODEL**

In this section, we explore how the Human Oracle Model can be used to describe the underlying algorithmic behavior and relative performance of human computation systems. The Image Labeling examples considered herein are canonical exemplars in the field; they are by no means a compre-hensive sampling of human computation in the wild. Instead, we leverage these simple examples and narrow use case in order to provide a concise vignette into the kinds of low-level evaluations and algorithmic comparisons that are made possible through the adoption of abstract models like the Human Oracle Model. In some cases, we have elected to model a slight variation of a system in the interest of facilitating a more interesting comparison. When this is the case, we will clearly document any modifications and provide justification for the change.

4.1. **The ESP Game**

The ESP Game[2] (von Ahn and Dabbish, 2004) is a human computation system designed to produce validated labels for images on the web. Each image is displayed to a randomly-assigned pair of

---

[2]Because of the widespread success of the original ESP Game and its many subsequent variants, many refer to the general collection of similarly incentivized input-agreement human computation systems as *ESP Games*. Because we will be exploring the relationships between several members of this larger class of

human collaborators, who are then asked to label the image in a finite amount of time. Because the human players cannot communicate with one another as they try to "agree" by guessing the same label, the dynamics of the game incentivize them to try guessing short, intuitive labels. A label is accepted once some number of pairs have agreed on it, and is then added to a list of TABOO words for that image. Future pairs are presented with the TABOO words in addition to the image, and these words are not accepted if guessed. This encourages the generation of new labels by future pairs, rather than simply repeating previously validated labels. For the purposes of this analysis, we will assume that just one pair must agree for a label to be accepted and that the computation continues until a match is found. The resulting output is a new description of the image that has been validated by both Human Oracles.

Recall that each Human Oracle has a finite list of labels they could use to describe a given image. In the best case, the algorithm terminates after only 2 queries, one to each Human Oracle whose first choice labels are a match. In the worst case, the Human Oracles' response lists A_{H_1} and A_{H_2} are exactly inverted and both have nonzero probabilities for all n possible labels. If this is the case, then each of the first n queries ($\frac{n}{2}$ to each Human Oracle) would result in a unique label before the $(n+1)^{st}$ query in forces a match. In the event that either H_1 or H_2 cannot return a new label, we assume that the computation will throw an error. When this occurs, the assumption that all pairs have a nonempty intersection in their response sequences allows us to infer that all valid labels for the input image are already listed in the TABOO list; if this were not the case, then the Human Oracles would have guessed the missing label.

4.2. KissKissBan

KissKissBan (Ho et al., 2009) is another human computation system designed to produce short, validated labels for images on the web. The authors note that "players tend to give easier and more generic descriptions [when playing the ESP Game], and therefore the diversity of the output becomes limited (Ho et al., 2009)." This system suggests an extension of the ESP Game intended to generate more creative labels by "introducing competitive element into the game (Ho et al., 2009)." Each image is shown to three online players. Two players are collaborating, as in the ESP Game, to try to guess the same label for the image. The other player, the Blocker, attempts to block the collaborative pair from matching by guessing a set of obvious labels at the onset of each round. These blocked labels are hidden from the collaborators, who are incentivized to match on a non-blocked word before their time runs out and are penalized for guessing blocked words. If they fail to match on a new word, the Blocker wins.

There are three ways a label can be validated during the game: (1) H_1's label matches one from $H_{Blocker}$, (2) H_2's label matches one from $H_{Blocker}$, or (3) H_1 and H_2 match on a label as in the ESP Game. Note that while matching on a blocked word produces a validated label, the game ends only on a match between the two collaborators. The resulting output is a *set* of labels that have each been validated by at least two Human Oracles.

We will presume that the Blocker generates $k-1$ labels at the onset of each game. In the minimal case, H_1 and H_2 match on their first label and this label is not blocked, requiring a total of $k+1$

---

systems, in this publication we will reserve the term "the ESP Game" to refer only to the original 2-player image labeling game published by von Ahn and Dabbish.

queries to generate a single label. Unlike with the ESP Game, the minimal case is not optimal in terms of minimizing queries-per-label. In the best case, the responses of H_1 and H_2 differ on first $k-1$ queries, but each response matches one of the $k-1$ blocked labels. They then match on their next guesses, requiring a total of $2k$ queries to generate k labels. In the worst case, the Blocker then responds to $k-1$ queries to generate the Blocked list, but none of these are matched by the collaborators. H_1 and H_2 are exactly inverted on their ordering of the remaining $(n-(k-1))$ responses, and the next query forces a match. When this is the case, KissKissBan requires $(k-1)+(n-(k-1))+1 = n+1$ queries to generate a single label.

4.3. **Comparing the ESP Game and KissKissBan**

Intuitively, the original implementation of the ESP Game and KissKissBan appear very similar both in terms of the problem they are trying to solve as well as the approach to finding a solution. Because their underlying problems are equivalent and their Human Oracles differ in number but not in function, we can directly compare the performance of the ESP Game and KissKissBan algorithms under the Human Oracle Model. We will begin by demonstrating that the worst case performance of the ESP Game requires no more queries per label than the worst case performance of KissKissBan.

Proof: Recall that in the worst case, KissKissBan returns just a single label with a large number of queries to the Human Oracles. All $k-1$ queries to $H_{Blocker}$ were wasted because none of the BLOCKED labels were matched, and the collaborators go $\frac{n-(k-1)}{2}$ rounds before finding a match for a total cost of $n+1$ queries. In this case, returning a single label could have been accomplished using one round of the ESP Game at an identical cost of $n+1$ queries. While the two Human Oracles may take just as long to find a match, there is no added benefit to including a third Human Oracle in the worst case. Thus, the worst-case number of queries to generate a single label in KissKissBan is equal to the worst-case cost of the ESP Game. ∎

We will next demonstrate that the best case performance of KissKissBan requires no fewer queries per label than the best case performance of the ESP Game.

Proof: In the best case, KissKissBan returns k unique labels using $2k$ queries to the Human Oracles: $(k-1)$ to $H_{Blocker}$ to set up the BLOCKED list, $(k-1)$ queries divided between H_1 and H_2, each of which matches a unique word on the BLOCKED list, and 2 final queries, one to each of H_1 and H_2, on which they match. This match causes the algorithm to terminate with a cost-per-label of 2. In the best case performance of the ESP Game, the pair is able to match on their first try for a total of 2 queries to the Human Oracles to generate a single label. Thus, the minimum number of queries per label in the best-case performance of KissKissBan is equal to the best case cost of k rounds of the ESP Game. ∎

From an algorithmic perspective, KissKissBan demonstrates no advantage over the ESP Game in terms of the **number of queries per label**. However, it is reasonable to argue (as do the authors) that KissKissBan (Ho et al., 2009) may produce "less obvious" labels than the ESP game in the short term. That is, the game dynamics in KissKissBan encourage the non-Blocking players to de-prioritize labels that they believe are likely to be guessed by the Blocker. More specifically, they are encouraged to avoid their initial top choices, assuming that those guesses will also be the top choices of their opponent. Despite this reordering, if played for enough rounds this model suggests there are no labels that KissKissBan would produce that would not also eventually be discovered in the ESP Game. Because both games are played using exactly the same kind of Human Oracle, the pool of potential labels is the same in both games; they may just be validated in a different order. This suggests that the difference in initial labels is due more to the incentive structure of the game than to any underlying computational differences, and challenges the claim that either algorithm provides *better* labels – they provide the same labels, in a different order. This is an important distinction: it enables us to isolate the cause of the effects we observe in the wild, and begin to explain the phenomenon in in computational terms. This comparison validates that the Human Oracle Model enables quantifiable comparison between algorithms that leverage the same affordance to solve the same problem.

4.4. **Peekaboom**

Peekaboom (von Ahn et al., 2006) is a human computation system designed to augment image labels with information about the location of the objects being described. Two players are partnered at random and are each assigned a role: Peek and Boom. Boom is presented with an image and a word, and Peek is presented with a blank screen. Boom is tasked with revealing just enough of the image to Peek so that she can guess the word. As Boom clicks on parts of the image, a small region of the image under the clicked location is revealed, and the incomplete image is sent to Peek.

The task given to Peek is identical to players of both the ESP Game and KissKissBan: given an image (in this case, an incomplete image), provide a description. Both players are incentivized to reveal and guess as efficiently as possible. The game infers that if Peek is able to guess the word, then Boom must have revealed the correct location. Once Peek has successfully matched the original word, a minimal bounding box is computed from the regions revealed by Boom. Experimental data suggest that the bounding boxes produced by multiple pairs when averaged tend toward minimally bounding the region containing the object. In this problem, we either assume that the textual description has been validated a priori, or that the computation will throw an error if the object does not appear in the image.

In the best case, one reveal from Boom is sufficient for Peek to guess correctly on the first try, for a total of 2 queries to validate the label. In the worst case, Boom must reveal all subregions of the entire $m \times m$ image before Peek can identify the correct label, resulting in a total of $O(m^2)$ queries to validate the label. In contrast to the two previous applications, in which humans are asked perform the same task and their responses are used to to verify one another, Peekaboom uses implicit validation on humans performing different tasks. This hints at a difference in the underlying computational machinery, which will be further discussed later in this work.

4.5. **Comparing the ESP Game and Peekaboom**

While both the ESP Game and Peekaboom compute on a single image and ask a a Human Oracle to describe the image as part of their computation, their underlying problems appear different. However, we can use reduction to demonstrate that the ESP Game can be used to solve the problem being solved in Peekaboom (bounding an object in an image).

Proof: Given an image I and a label describing an object to bound, nondeterministically select a subimage I'. On I', iterate the ESP Game to return all possible descriptions. If any of the returned

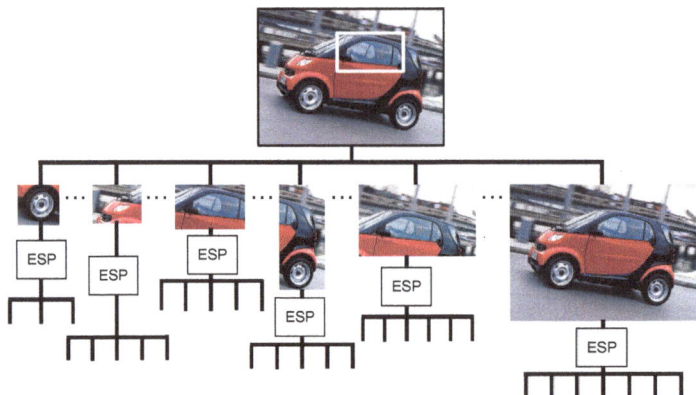

Figure 1. Bounding an object labeled* driver *using iterative queries to the ESP Game on subimages.

labels matches the input label, return the boundary of the subimage as the bounding box. ■

The number of possible subimages is limited by the size of the image. As before, the number of possible valid descriptions for any input image is also finite due to the limitations of both image resolution and language. Thus, if the label is valid for the subimage, it will eventually show up as one of the suggested descriptions returned by the ESP Game. We are therefore guaranteed that this nondeteriministic "guess-and-check" method will eventually yield a correct bounding box. This reduction is depicted graphically in Fig. 1.

We now demonstrate that the maximum number of queries to the Human Oracle using either Peekaboom or repeated rounds of the ESP Game required to bound an image are both polynomially bounded in the size of the image, these polynomial bounds are not equivalent.

Proof: Recall that the goal of Peekaboom is to return a minimal $w \times h$ bounding box surrounding the described object in the image, and that this is accomplished by having Boom sequentially reveal parts of the image to Peek. Assume without loss of generality that the size of the image is $m \times m$, and that the size of each revealed region is $r \times r$, where $0 < r < m$. The smallest possible bounding box, where $w = h = r$, would be returned in the case that Peek was able to guess the word after a single reveal. In the worst case $w = h = m$, because Peek may not be able to guess the word before seeing the entire image, which could require at most $2 * (\frac{m}{r})^2 = O(m^2)$ queries to the Human Oracles.

As indicated above, we can repeatedly invoke the Oracle-ESP algorithm on each subimage in ascending order of size until either the algorithm returns a matching label or we exhaust the possible labels for the subimage without finding a match, and move on to the next one. Because (1) the label given as input to Peekaboom was validated a priori, (2) there are finitely many valid labels for any image, and (3) the Oracle-ESP algorithm will eventually return all valid labels, we can be assured that this process eventually terminates. Because we evaluate subimages in increasing order of size, this process guarantees that the first subimage on which we find a match is minimal.

The total number of times we must play the ESP Game is determined by the number of subimages that must be evaluated. The smallest bounding box that could be returned by Peekaboom is the size of one revealed region, and so we need only evaluate subimages that are at least $r \times r$, and that are at most the entire image. The number of possible subimages ranging from size $r \times r$ to $m \times m$ is:

$$\sum_{w=r}^{m} \sum_{h=r}^{m} (m-w+1)(m-h+1) = O(m^4)$$

thus requiring on the order of $O(m^4)$ queries to the Human Oracles across all executions of the algorithm. Thus, the worst-case number of queries needed to bound an object using only the ESP Game grows asymptotically faster than the number of queries needed using Peekaboom. ∎

In the proofs above, we used brute force to illustrate the relationship between the ESP Game and Peekaboom. This demonstrates the relationship between *labeling an image* and *locating an object in an image* in an intuitive manner, but we reiterate that this is not intended as a prescription for future design. In practice, because this method requires an exhaustive search, this approach would not be an effective use of human computation resources. The average case performance could likely be significantly improved by making more intelligent decisions about which subimages to send to the subroutine for validation. We could, for example, start with a well-distributed subset of subimages of each size. This has the potential to greatly reduce the number of calls to the subroutine because it could de-prioritize redundant rounds on uninteresting regions of the image without sacrificing accuracy. We could also select regions according to content by preprocessing using an image segmentation algorithm. This would increase the amount of machine computation, in exchange for a reduction in the number of queries to the Human Oracles. However, these heuristics would not alter the underlying differences between these algorithms.

5. IMAGE VS. AUDIO LABELING

In the above comparisons, we demonstrated that the ESP Game can be used to approximate several other human computation algorithms. However, this is not meant to suggest that the ESP Game is somehow universal or complete. For example, consider whether a successful image labeling algorithm such as the ESP Game (von Ahn and Dabbish, 2004) can be used to label other stimuli such as audio. Such a reapplication was attempted by the original designers to create a system called TagATune (Law et al., 2007). Despite being identical to the ESP Game in nearly every way with the exception of the human affordance, this first iteration failed miserably; people simply couldn't agree on labels for most of the input. In a second iteration, the designers found that asking the users to decide whether or not they thought they were listening to the same thing was far more successful for audio labeling than explicit matching (Law and von Ahn, 2009), although this introduces significantly more noise into the resulting dataset.

This would indicate that though the human is superficially being asked to perform a similar task, the underlying information processing is fundamentally different for images versus audio. Our lived experience might lead us to speculate that the human might be sampling their responses from a much larger search space. That is, perhaps the sets $R_Q(x)$ from which Human Oracles are drawing their responses are dramatically larger for audio inputs than for images. This makes intuitive sense;

after all, audio lacks the same tangible, concrete concepts like *chair* and *grass* upon which we often anchor our labels for images. In addition, one might suggest that the fact that the input is continuous rather than discrete might play some role. Whatever the underlying reason, this suggests that in some sense audio labeling is a *harder problem* than image labeling, and that it therefore requires more powerful machinery to solve. In the following section, we define several dimensions which can be used to compare the relative strength of systems leveraging human computation and demonstrate that these dimensions define preliminary complexity classes on the space of human computation.

6.   DIMENSIONS AND COMPLEXITY CLASSES

As demonstrated in the previous section, the number of required operations is one intuitive metric by which we may order a collection of algorithms for solving a problem. Indeed, this is analogous to traditional notions of computational work. Because we lack a mechanism for converting between units of *human work* and units of *machine work*, the $\langle \Phi_H, \Phi_M \rangle$ notation introduced by Shahaf and Amir (Shahaf and Amir, 2007) can prove useful.

Recall for example the techniques discussed previously for identifying the location of an object in an $m \times m$ image. Using Peekaboom, the number of queries to the Human Oracles is bounded by $O(m^2)$. The cost to the machine is a simple comparison between each of Peek's guesses and the input label, and so $\Phi_M = \Phi_H = O(m^2)$ as well. In the second approach using the ESP Game as a subroutine, the number of queries to the Human Oracles could be as high as $O(n^2 * m^4)$ in the event that all n labels need to be validated for each subimage before we arrive at the input label. The machine must then compare the value returned by each query to the collection of previous guesses for that round to determine if there has been a match. Assuming an efficient data structure, each lookup or insertion would incur a cost on the order of $\log(n)$. In addition, each of the n possible returned labels must be compared against the input label. This results in a total machine cost of $O(n^2 * m^4 * \log(n) + n)$. Comparing these two tuples, it is clear that Peekaboom is a more efficient method for bounding an object in an image than the ESP Game approximation in terms of both human and machine computation.

Perhaps more interestingly, the examples given demonstrate that an algorithm requiring **more information** (Peekaboom requires a predefined label) as well as **interactivity** (Boom must be able to touch the image) can be simulated using a polynomial number of calls to an algorithm with limited interactivity and unrestricted search space (ESP). This sheds important light on the *value* of this additional information and power, and the scale of the cost incurred if we are forced to solve the same problem using an intuitively "weaker" machine.

This notion of "stronger" and "weaker" suggests that the way in which human computation is leveraged as part of an algorithmic process may be used to induce *complexity classes* that partition the space of human computation. By developing an intuition about the ways in which problems group together in terms of how their solutions leverage human computation as a resource, we can better equip ourselves to exploit what we already know to solve novel problems. In the following sections, we will discuss several additional dimensions along which human computation may be classified and compared.

6.1. Query Order

First, we can consider whether or not the sequence of queries to the Human Oracle can be determined in advance. In an algorithm with **predetermined query order**, we claim that there exists some function:

$$f : I \to (q_1, \ldots, q_n)$$

that takes as input a problem instance I and generates a finite sequence of queries $(q_1, \ldots, q_n)$ that will be passed in order to the Human Oracle. Because the sequence can be generated *a priori*, it follows that the position of any query q_i must not depend on the Human Oracle's previous answers. In these systems, the Human Oracle cannot influence the order in which it receives and responds to specific queries. A general example of a process that uses predetermined queries is semi-supervised machine learning. In these techniques, the Human Oracle is asked to label a set of training data which is then used to infer a classification function. While the resulting classification function is dependent on how the training data is labeled, the points to be used as training data are determined in advance.

Relaxing this restriction yields algorithms whose future queries may be contingent on the Human Oracle's answers to previous queries. In an algorithm with **adaptive query order**, we claim that there exists some function:

$$
\begin{aligned}
f : \{I, \emptyset\} &\to q_1 \\
f : \{I, (a_1)\} &\to q_2 \\
&\vdots \\
f : \{I, (a_1, \ldots, a_n)\} &\to q_{n+1}
\end{aligned}
$$

that takes as input a problem instance I as well as $(a_1, \ldots, a_n)$, the sequence of responses from the Human Oracle so far, and generates the next query (q_{n+1}). An excellent example of adaptive querying is *active learning*. In active learning algorithms, the Human Oracle is first asked to label some small set of training data. Based on their responses, the algorithm reclusters the data. It then selects a new subset of points about which it is least confident and submits these to the Human Oracle to label. The selection of each subsequent collection of points is directly related of the labels provided in the previous round. This process continues iteratively until some confidence threshold is met.

Any result that can be computed on a machine with predetermined query order can also be computed on a machine with adaptive query order; the adaptation would simply be to do nothing with the additional information. The inverse is not true, as the decisions made by an adaptive query machine regarding which question to ask next may not be predeterminable. Thus, the class of problems solvable using a predetermined query machine is contained within the class of problems solvable using an adaptive query machine (Fig. 2).

6.2. Oracle Responses

Mirroring the previous dimension, some systems may treat the Human Oracle's responses to queries as largely **independent** of one another. One example of a real-world system where this is true is

Figure 2. Illustration of the relationship between adaptive and predetermined query order machines. In a predetermined query order, all queries to the Human Oracle must be determined a priori;. In an adaptive query order, subsequent queries to the Human Oracle may be adapted based on the response to previous queries. Any result that can be computed on a machine with predetermined query order (left) can also be computed on a machine with adaptive query order (right).

reCAPTCHA (von Ahn et al., 2008), a human computation system for optical character recognition (OCR). If the same human is asked to pass several reCAPTCHA instances, any response bias due to priming or learning would likely be negligible. Thus, for analytical purposes, we can presume that each of her responses is independent. In practice, processes leveraging such sequence-independent responses may be parallelizable or perhaps more robust to interruption or participant turnover. In such cases, there would be no discernible difference between asking 10 Human Oracles to each answer 10 queries, or asking 100 different Human Oracles 1 query each. Because these Human Oracles have no practical awareness of their previous responses, we refer to them as *forgetful.*

It is sometimes useful to endow the Human Oracle with some amount of persistent memory regarding the query sequence. In these systems, the Human Oracle may be able to **modify its future behavior based on previous events**. In the simplest sense, this could be used to ensure that the Human Oracle does not return the same answer twice as in the previous examples. In other scenarios, we may wish to enable computation on this sequence in order to model more nuanced adaptations, such as learning or fatigue. For example, complex systems such as visual analytics tools require that the Human Oracle be able to learn from a prior sequence of actions and responses and subsequently modify its future behavior. Note that while we continue to develop more robust models of human memory and its limits, we may abstract the specifics of *how* the Human Oracle remembers and instead include the cost of accessing this memory in the query cost. Because these Human Oracles can remember and compute on their previous answers, effectively gaining knowledge from their response history, we refer to them as *wise.*

As with the previous dimension, any result that can be computed on a machine with a forgetful Human Oracle can also be computed on a machine with a wise Human Oracle; when simulating a forgetful machine, the wise Human Oracle's memory remains unused. The inverse is not true, as the responses returned by a wise Human Oracle may not be parallelizable. Thus, the class of problems solvable using a forgetful Human Oracle is contained within the class of problems solvable using a

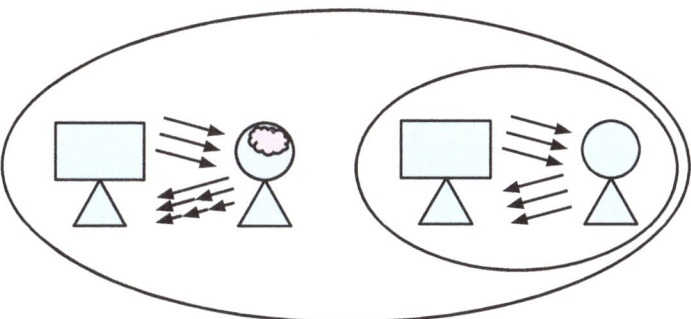

Figure 3.** **Illustration of the relationship between machines that use wise and forgetful Human Oracles. A forgetful Human Oracle does not gain any information from answering previous queries; a wise Human Oracle may use previous queries to influence its responses in the future. Any result that can be computed using a forgetful Human Oracle (right) can also be computed using a wise Human Oracle (left).

wise Human Oracle (Fig. 3).

6.3. **Supplemental Information**

We can also consider whether or not the Human Oracle can inject supplemental information into the computation process on its own. This is tantamount to asking whether or not the Human Oracle is allowed to provide answers to questions that were not directly asked. In many existing human computation algorithms, the human or collection of humans is asked to perform a *specific* computational process as a subroutine to a larger algorithm, such as labeling a particular image. Under this restriction, the Human Oracle does not have any power to interject new information or redirect the computational process. This is consistent with the standard Oracle Turing Machine model; the Oracle is simply a set, and can only answer questions about membership in that set. The algorithms under consideration in this work and in many other *Games with a Purpose*-style systems, such as Fold.it (Cooper et al., 2010) for finding protein foldings, fall within this category.

Alternatively, the Human Oracle may be given some autonomy regarding the interjection of supplemental information into the computational process. That is, in addition to responding to queries generated by the computational process, the Human Oracle may spontaneously provide new information that may then impact future computation. This adds an additional level of computational power to the Human Oracle. For example, the Human Oracle may have its own computational agenda such as in the use of Visual Analytics systems, and may therefore inject supplemental information to direct the computation in a specific direction. Indeed, we suggest that this supplemental information may in fact be integral to the success of many Visual Analytics systems that leverage the domain expertise or lived experience that is yet impossible to simulate within purely mechanical analysis.

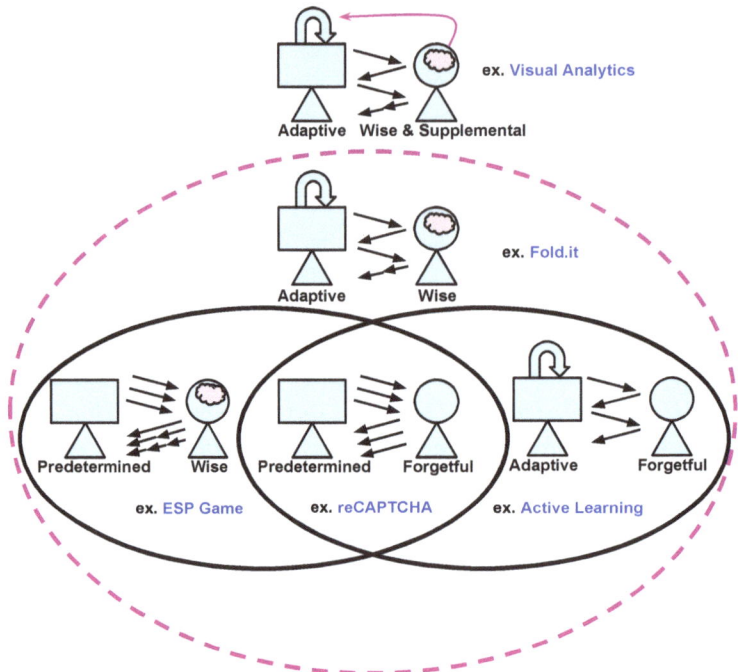

Figure 4. ***Preliminary hierarchy of complexity classes in human computation defined under the Human Oracle Model.***

7.   COMPARING HUMAN COMPUTATION SYSTEMS IN COMPLEXITY SPACE

In contrast to previous schema for comparing human computation systems which rely on nominal classification dimensions (Quinn and Bederson, 2011), each of the dimensions introduced here has an implicit notion of magnitude that induces a partial ordering on different algorithms and problems. Categorizing along these three dimensions, many of our intuitions about the relative strength of existing techniques are captured. For example, the algorithm underlying reCAPTCHA can be computed using a less powerful Human Oracle than the ESP Game. In reCAPTCHA, the human is simply a visual decoder, and the machine takes care of compiling and aggregating the results. In the ESP Game, more responsibility is placed on each individual human to avoid performing redundant computation in order to generate a selection of unique labels. Similarly, we see that active learning requires a more powerful use of human computation than semi-supervised learning. We presume that by enabling more careful selection of user constraints or labels, the set of datasets that can be classified using active learning is broader than the set of datasets that could be classified using semi-supervised learning, given the same amount of supervision.

Because they define a partial ordering between problems, these dimensions can be used to establish preliminary *complexity classes* within the space of human computation (Fig. 4). We suggest that these classes are complementary to those in traditional computational complexity. Indeed, we may consider these hierarchies to be orthogonal. That is, there could exist human-computer collaborative systems with all combinations of $\langle \Phi_H, \Phi_M \rangle$. For example, the ESP Game lives at the intersection of *predetermined query order, wise Human Oracle* and $O(n^2) \in$ P, while Fold.it exists where *adaptive query order, wise Human Oracle* meets NP.

This hierarchy is still a work in progress; many of the relationships between these classes are still unknown (Fig. 5). For example, how does the class of problems solvable by a *fixed query order, forgetful Human Oracle* machine with *supplemental information* relate to the class of problems solvable using a machine with *adaptive query order, a forgetful Human Oracle* and *no supplemental information*? How does this compare to a machine with *fixed query order, a wise Human Oracle* and *no supplemental information*? These questions, along with the discovery of canonical or **complete** problems in this space, are the subject of our future research in this area.

8. DISCUSSION

The Human Oracle Model provides a critical first step in quantifying human computation, and helps us to better understand the intricate relationships among different problems and problem families when viewed through the lens of human computation. That said, this work only just scratches the surface of this potentially rich area for future research. To start, this paper considers only boundary cases: that is, what are the upper and lower bounds on the work that is being done? While it is beyond the scope of this initial exercise, it would also be very informative to assess the average case performance of these algorithms as well.

In addition, this model ignores some very real factors present in any system involving the variability of biological computation. In the following sections, we discuss some of the limitations of this model, as well as motivate continued research in this area.

8.1. Imperfect Oracles

Under this model, there is an explicit assumption the Human Oracle will always be able to provide the correct answer at a fixed cost. In reality, humans don't work this way. Intuition and experience indicate that humans eventually get tired or bored, and as a consequence their speed and accuracy suffer. In addition, individual differences in ability and cost are absent. In the real world, not all humans are equal in their capacity to answer the questions we ask. Some are more skilled or have better training, and their expertise comes (we presume) at a higher cost.

Similar issues have arisen in the area of active learning, which has historically assumed a single tireless, flawless, benevolent Oracle was always available to provide labels for its training data.

Figure 5. Open question: how does a **fixed query order, forgetful Human Oracle** *machine with* **supplemental information** *(center) relate to the class of problems solvable using a* **fixed query order, wise Human Oracle** *machine with* **supplemental information** *(left) or an* **adaptive query order, forgetful Human Oracle** *machine with* **no supplemental information** *(right)?*

Proactive learning relaxes these assumptions, adopting a decision-theoretic approach to match one of a collection of (possibly imperfect) Oracles to each instance (Donmez and Carbonell, 2008). More recent work in the area of *multiple expert active learning* (MEAL) improves upon this model by incorporating load balancing to ensure that no worker has to shoulder an inequitable share of the burden (Wallace et al., 2011). These methods assume there exists some method to model both how hard any single problem instance is, as well as how costly and effective a given worker is.

8.2. Quantifying the Human Brain

This highlights another problem: as of this writing, there does not exist any reliable method for quantifying how hard a human has to work in order to accomplish a given task. Because we don't fully understand the fundamental operations of the human brain or how they assemble to perform computation, it is not yet possible to calculate a precise per-operation cost. As such, at present this model cannot actually tell us *how much work* the human is doing; it only tells us how many times the human is working (query complexity). When the task is comparable, such as when comparing various image labeling algorithms, this does not pose a significant problem. But what about comparing dissimilar processes?

While cognitive modeling techniques can help us to understand the interplay between stimulus and response, existing architectures are not designed to determine the "complexity" of the model itself. Unobtrusive brain sensing methods are currently under development and have shown promise in detecting task difficulty (Girouard et al., 2009), but the information revealed is not yet refined to a per-operation granularity. Thus, from a cognitive science perspective, there is presently no mechanism for quantifying the computation performed by the human brain. In order to form a complete model of human computation, it is critical that we continue to develop more nuanced models of the human brain and to incorporate these models into the evaluation of algorithmic complexity and performance in human-machine collaborative systems.

9. CONCLUSION

The importance of understanding human computation as part of a larger computational complexity system is not limited to improving algorithm design. Augmenting computational complexity models to incorporate human computation can expand our understanding of what can be computed, as did the development of probabilistic and parallel computation. The development of complexity measures for human computation may play a significant role in the broader adoption of human computational methods. Robust models of *how humans fit* into the grand scheme of computational tools is essential to promoting wider understanding of human effort as a legitimate and measurable computational resource.

In this work, we introduced the Human Oracle Model as a method for characterizing and quantifying the *use of human processing power as part of an algorithmic process*. We demonstrated the utility of this model for comparing and analyzing several well-known human computation systems for image labeling and described how this model can be used to characterize the space of human computation. In closing, we discussed the model's limitations and its potential for broader impact. Through this research, we hope to form a more holistic picture of the interrelationship between human and machine computation, and to develop a robust theoretical model for the analysis of systems involving their collaboration.

10. REFERENCES

Arora, S and Barak, B. (2009). *Computational complexity: a modern approach.* Cambridge University Press.

Bertini, E and Lalanne, D. (2010). Investigating and reflecting on the integration of automatic data analysis and visualization in knowledge discovery. *SIGKDD Explorations Newsletter* 11, 2 (2010), 9–18.

Cooper, S, Khatib, F, Treuille, A, Barbero, J, Lee, J, Beenen, M, Leaver-Fay, A, Baker, D, Popovic, Z, and others, . (2010). Predicting protein structures with a multiplayer online game. *Nature* 466, 7307 (2010), 756–760.

Crouser, R and Chang, R. (2012). An Affordance-Based Framework for Human Computation and Human-Computer Collaboration. *Visualization and Computer Graphics, IEEE Trans. on* 18, 12 (2012), 2859–2868.

Donmez, P and Carbonell, J. (2008). Proactive learning: cost-sensitive active learning with multiple imperfect oracles. In *Proc. 17th ACM Conf. on Information and knowledge management.* ACM, 619–628.

Girouard, A, Solovey, E, Hirshfield, L, Chauncey, K, Sassaroli, A, Fantini, S, and Jacob, R. (2009). Distinguishing difficulty levels with non-invasive brain activity measurements. In *Human-Computer Interaction.* Springer, 440–452.

Ho, C, Chang, T, Lee, J, Hsu, J, and Chen, K. (2009). KissKissBan: a competitive human computation game for image annotation. In *Proc. SIGKDD Workshop on Human Computation.* ACM, 11–14.

Law, E and von Ahn, L. (2009). Input-agreement: a new mechanism for collecting data using human computation games. In *Proc. 27th SIGCHI Conf. on Human factors in computing systems.* 1197–1206.

Law, E, von Ahn, L, Dannenberg, R, and Crawford, M. (2007). Tagatune: A game for music and sound annotation. *Proc. of ISMIR (Vienna, Austria)* (2007).

Lawler, E. L, Lenstra, J. K, Kan, A. R, and Shmoys, D. B. (1985). *The traveling salesman problem: a guided tour of combinatorial optimization.* Vol. 3. Wiley New York.

MacGregor, J. N and Ormerod, T. (1996). Human performance on the traveling salesman problem. *Perception & Psychophysics* 58, 4 (1996), 527–539.

Mason, W and Watts, D. (2010). Financial incentives and the performance of crowds. *ACM SigKDD Explorations Newsletter* 11, 2 (2010), 100–108.

Quinn, A and Bederson, B. (2011). Human computation: a survey and taxonomy of a growing field. In *Proc. 29th SIGCHI Conf. on Human factors in computing systems.* ACM, 1403–1412.

Shahaf, D and Amir, E. (2007). Towards a theory of AI completeness.. In *Logical Formalizations of Commonsense Reasoning.* 150–155.

Singer, Y and Mittal, M. (2011). Pricing Tasks in Online Labor Markets.. In *Proc. Workshop on Human Computation at the 25th AAAI Conf. on AI.*

Turing, A. (1938). *Systems of logic based on ordinals: a dissertation.* Ph.D. Dissertation. Cambridge.

von Ahn, L. (2006). Games with a purpose. *Computer* 39, 6 (2006), 92–94.

von Ahn, L and Dabbish, L. (2004). Labeling images with a computer game. In *Proc. 22nd SIGCHI Conf. on Human factors in computing systems.* ACM, 319–326.

von Ahn, L, Liu, R, and Blum, M. (2006). Peekaboom: a game for locating objects in images. In *Proc. 24th SIGCHI Conf. on Human factors in computing systems.* ACM, 55–64.

von Ahn, L, Maurer, B, McMillen, C, Abraham, D, and Blum, M. (2008). reCAPTCHA: human-based character recognition via web security measures. *Science* 321, 5895 (2008), 1465–1468.

Wallace, B, Small, K, Brodley, C, and Trikalinos, T. (2011). Who Should Label What? Instance Allocation in Multiple Expert Active Learning.. In *SDM.* 176–187.

Human Computation (2014) 1:1:67-93
ISSN: 2330-8001, DOI: 10.15346/hc.v1i1.5

Architecting Real-Time Crowd-Powered Systems

WALTER S. LASECKI, UNIVERSITY OF ROCHESTER

CHRISTOPHER HOMAN, ROCHESTER INSTITUTE OF TECHNOLOGY

JEFFREY P. BIGHAM, CARNEGIE MELLON UNIVERSITY

ABSTRACT

Human computation allows computer systems to leverage human intelligence in computational processes. While it has primarily been used for tasks that are not time-sensitive, recent systems use crowdsourcing to get on-demand, real-time, and even interactive results. In this paper, we present techniques for building real-time crowdsourcing systems, and then discuss how and when to use them. Our goal is to provide system builders with the tools and insights they need to replicate the success of modern systems in order to further explore this new space.

---

1. INTRODUCTION

Crowdsourcing elicits responses from groups of people via an open call to action (Surowiecki, 2005). Such calls have recently begun being used to elicit people's contribution to human computation processes (von Ahn, 2005). This work might be done for pay, for a sense of accomplishment, or because the worker gets some direct benefit from the final result. Human computation allows systems to go beyond what automated approaches alone are capable of and, using large groups, leverages the "wisdom of crowds" to solve problems that cannot be easily solved by an individual.

In the past, the time needed to both recruit crowds and complete tasks meant that crowdsourcing was primarily an offline mechanism, where the completion time was a low priority. However, more recent research has yielded approaches that leverage the scale and structure of current platforms, such as Amazon Mechanical Turk, to collect answers in seconds, not hours or days (Bernstein et al., 2011; Bigham et al., 2010). This allows task requesters to leverage crowds on demand for interactive applications (Lasecki et al., 2011), e.g., converting speech to text fast enough to provide deaf users with access to their environment (Lasecki et al., 2012). The robustness of these systems also makes it possible to deploy them in real settings, and use the collected data to train automated systems to play increasingly larger roles (Lasecki et al., 2013a).

However, since these systems require the understanding of a wide variety of issues – ranging from

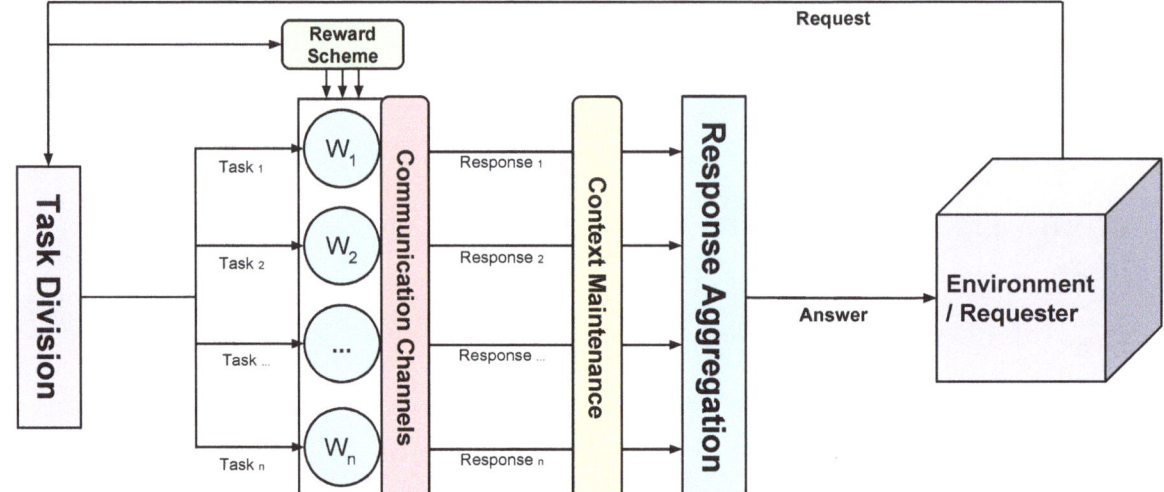

Figure 1. *Our real-time crowd-powered system architecture.*

interface design to human factors to game theory to aggregation and voting algorithms – the amount of broad and specialized knowledge and experience needed to build them is a major barrier.

Our goal is to inform the design and implementation of real-time crowd-powered systems. We strive to make the problem of integrating real-time crowdsourcing into interactive systems more approachable for system builders by reducing the need for significant prior domain expertise in crowdsourcing in order to develop solutions. We focus on the tools, methods, and paradigms that have been successful in prior work, and discuss when and how different components and approaches should be used in order to be most successful. We also consider recurring patterns of classes and communicating objects that recur in real-time crowdsourcing. This information has not previously been available in a general, interconnected way, since work on most prior systems has been separated into task-focused systems and studies.

In the rest of this paper, we will discuss the following aspects of real-time crowd systems of crowd-powered systems (this architecture is visualized in Figure 1):

– Task division
– Answer aggregation
– Maintaining context
– Facilitating communication between workers
– Rewarding workers
– Integrating automated systems

2.   BACKGROUND AND RELATED WORK

Using crowdsourcing for human computation requires ideas from a wide range of fields, from human factors and human-computer interactions to decision science and voting theory. In this section, we briefly discuss some of the foundational work tying these fields to crowdsourcing.

2.1. Definitions

We begin by defining some of the terms as we use them in this paper:

- **Task:** A job or computation that requires human input to complete.
- **Requester:** A person or system who needs an answer to a *task* from the crowd.
- **Crowd:** A dynamic group of workers available via an online platform to complete tasks.
- **Crowd Worker ("worker"):** A person completing tasks on a crowdsourcing platform.
- **Microtask:** A small, context-free task that is common in crowdsourcing.

2.2. Related Work

Crowdsourcing spans many fields: collective intelligence, social computing, economics, and human-computer interaction just to name a few. Collective intelligence studies how the dynamics of groups leads to intelligent and emergent behaviors, and social computing studies how computationally mediated interactions between people arise and their effects on users. It also encompasses the design of purpose-built systems that facilitate and moderate interactions. Human-Computer Interaction (HCI) and other Computer Science disciplines have mainly focused on crowds of workers than can be recruited algorithmically (Little et al., 2010; Bernstein et al., 2010, 2011; Bigham et al., 2010; Lasecki et al., 2012). This work focuses on crowd-powered systems that are able to algorithmically recruit and organize workers for a specific, task-focused objective. A discussion of crowdsourcing and related disciplines can be found in (Quinn and Bederson, 2011) and (Michelucci, 2013).

There are a number of crowdsourcing challenges we explore in this paper, including:

- **Dynamic workforces and worker availability.** One of the greatest benefits of participating in crowd work as a worker is the flexibility to take tasks when it is convenient for them, and stop working when they choose to. However, for requesters, this flexibility means that they cannot rely on particular workers to be available to complete a task when needed, or for workers to remain present over the course of multiple tasks (Mao et al., 2013). Even within a single task, workers may leave if the need arises or simply if they do not enjoy or understand the task.
- **Variable system configurations.** The stability and configuration of worker computers is not as controlled as in a traditional work setting. The setup and capabilities of workers' computers can vary widely even within the same platform. This means that task design must carefully consider what technologies might prevent some workers from participating in a task. Computer glitches can also result in workers disconnecting from a task before completion.
- **Answer reliability.** The crowd's anonymity (leading to minimal accountability), situational and experiential variability (leading to differing worker performance between tasks), and the lack of a strong, quality signal (leading to the requester having no reliable way of vetting workers prior to hiring them), all result in a process that is often unwieldy and potentially unreliable.

It is important to note that some of these issues (those involving worker reliability and availability) arise more frequently when using general-knowledge crowd marketplaces (such as Mechanical Turk), where specific skills are not on display. Longer term interactions are commonly seen on expert crowdsourcing platforms, such as oDesk (https://www.odesk.com/). As general knowledge platforms evolve and grow, it is likely that many of these properties will tend towards those seen in more expert platforms (Kittur et al., 2013).

2.3. Crowd-Powered Systems

The ability to correct worker input and get a reliable signal makes it feasible to create task-oriented systems powered by the crowd. In this section, we outline some of the work that has led from reliable offline answers from the crowd to systems that can interact with users in real time.

2.3.1. Making Crowds Available On-Demand

Being able to respond quickly, often in one second or less, is a key component of most interactive systems (Bernstein et al., 2011). However, when using people to power the system's responses, reducing latency becomes difficult. Worse, if these workers must be recruited each time an interaction might occur, the problem of reacting in a suitable amount of time becomes difficult.

To address the shortcomings of these systems in interactive settings, some systems have explored methods for ensuring that workers are always available for a task as soon as they are needed. VizWiz (Bigham et al., 2010) recruits workers in advance and keeps them engaged by completing prior tasks in order to train workers while keeping them available when needed. (ready as soon as their most recent task is done). Adrenaline (Bernstein et al., 2011) used a passive waiting approach combined with pop-up alerts to keep people available, not busy completing a separate task. Instead, multiple workers were recruited and asked to keep a browser window open which alerts them as soon as a task arrives. By recruiting a larger crowd, the fastest-to-respond member can be recruited, speeding the response time up even more (Lasecki et al., 2013). By integrating queuing theory and other optimizations involving the size of the potential workforce, it is possible to get workers to join a task in less than a second (Bernstein et al., 2012).

2.3.2. Synchronous Crowds

Recruiting individual workers quickly is important for making crowd-powered systems feasible in interactive settings, but does not allow for many of the consensus-based approaches common in crowd-powered systems that were discussed above. To enable this type of quality assurance process, we need to recruit multiple workers in real-time, often *synchronously*.

Synchronous crowd are necessary for systems in which workers build on one another's responses to generate a single answer. For example, Adrenaline (Bernstein et al., 2011) asked workers to rapidly refine others' cropped video segments until a single best still image was selected, while other work has used synchronous pairs of workers to collect task-oriented text dialogs using crowd workers (Lasecki et al., 2013a).

Finally, there is a growing body of research on the effects of different task coordination schemes on overall system performance. For example, Seaweed (Chilton et al., 2009) had Mechanical Turk workers play economic games, creating an easy platform for economists to study models with real people. Similarly, TurkServer allows social scientists to run synchronous experiments on Mechanical Turk (Mao et al., 2012), which has previously been shown to be a valuable resource for offline tasks (Mason and Suri, 2012).

2.4. **Real-Time Crowd-Powered Systems**

Being able to recruit workers on demand, to work synchronously on a task, makes it possible to create systems that interact with end users in a way that artificial intelligence alone cannot currently support. VizWiz (Bigham et al., 2010) was one of the first such systems – it allowed blind users to take a picture, ask a verbal question about the picture, and get an answer in under 30 seconds. Adrenaline (Bernstein et al., 2011) allowed users to find the best still image in a short video in a matter of seconds, allowing end users to take better pictures with almost no additional effort.

Legion (Lasecki et al., 2011) was one of the first systems to be able to continuously react to its environment as fast as a person could. Legion allowed users to control any existing user interface (UI) with natural language by having the crowd collectively control the UI remotely. This enabled off-the-shelf systems ranging from word processors to robots to be controlled using natural language, without needing any custom applications per-tool. Legion was the first to introduce the *crowd agent* model of crowdsourcing, where the goal is to coordinate the crowd to act as a single collective individual.

Chorus (Lasecki et al., 2013) went beyond natural language control using single commands, and let users easily find information via a conversational assistant that can hold extended dialog with the end user. Chorus asked workers to build on, refine, and vote for one another's conversational responses to create an intelligent dialog system that provides more accurate and thorough responses than any one worker alone would. Chorus:View (Lasecki et al., 2013b) used the Chorus conversational platform to extend VizWiz into a conversational visual question-answering assistant capable of giving real-time feedback to users regarding how to accurately frame their image.

In this paper, we explore the most successful approaches to mediating crowds for interactive tasks, outline and identify the most-appropriate usage scenarios for different components and system designs, and report new approaches and negative results that have not been reported in prior work. This allows system builders to replicate the success of modern interactive crowd-powered systems by appropriately applying these techniques to their projects.

2.5. **Applications of Real-Time Crowd-Powered Systems**

Real-time crowd-powered systems can have an important impact on many areas, include machine learning and accessibility.

2.5.1. *Machine Learning*

One of the earliest uses of crowdsourcing was to train automated systems. Image labeling (Russell et al., 2008; von Ahn and Dabbish, 2004), natural language tagging (Snow et al., 2008), language translation (Callison-Burch, 2009), and image sketching (Limpaecher et al., 2013) were among the many settings where crowd input was used to train machines. However, like other crowd tasks, these systems got responses from the crowd in an offline, batch-processing fashion.

Real-time crowdsourcing systems, make it possible to supplement and train the system in-situ by getting just-in-time answers from the crowd. For example, Legion:AR (Lasecki et al., 2013a) provided real-time activity recognition labels to train an Hidden Markov Model when the system was not confident in its own answer. Legion:AR could provide training as needed, and hide the system's training process from the user by providing reliable, direct answers.

2.5.2. Accessibility

Crowd-powered systems have been used to support accessibility since early systems. In fact, remote human-powered support of people with disabilities has long been used, even before the rise of modern crowdsourcing (Bigham and Ladner, 2011).

Interactive systems such as VizWiz, Chorus:View, and Legion:AR, have stretched the types of services that can be provided by these systems farther than ever before. Many of these services would not be possible to provide without a dynamic workforce such as online crowds provide. Others can be provided, but are either not as available to end users, or might even not be as high quality as the crowdsourced solution. For instance, Scribe (Lasecki et al., 2012) is a system that allows deaf and hard of hearing users to have better access to the world by providing real-time captions with a latency of under 5 seconds on demand.

3.   TYPES OF CROWDS

There are many different ways to recruit groups of workers to a task, and each comes with a trade-off in terms of their motivations, error rate, abilities, cost, and difficulty to recruit. Selecting a type of crowd is a key issue that arises in any crowd-powered system, but the differences can be particularly pronounced in real-time systems. Prior work has analyzed specific platforms for demographic break-downs of their members, which provides insight into education and engagement on the platform (Ipeirotis, 2010). In this section, we broadly outline the types of crowds that exist, their properties, and the pitfalls and opportunities that are associated with each.

3.1.   Source of the Crowd

The type of call, as well as the population targeted by the call, play one of the most significant factors in guiding the design of the rest of the system.

Open crowds, as the name implies, are open to any (or many) contributor(s). In practice, these workers are often recruited as paid microtask workers on platforms such as Mechanical Turk, or recruited as volunteers interested in the final result of the system (e.g., supporting accessibility systems (Bigham et al., 2010; Lasecki et al., 2013b, 2012)). Due to its position as the most ubiquitously used crowd platform in the crowdsourcing literature, and the fact that it contains many examples of the types of pitfalls we try to address in this paper, we use Mechanical Turk as our default platform for examples, unless explicitly mentioned otherwise.

Known crowds are crowds whose members are either known to the requester (friends, associates, full-time employees, etc.), or are otherwise known not to be malicious. Knowing that workers can be trusted – even if they cannot be relied upon to have any special skill set – can make a significant difference in choices made during implementation, particularly when it comes to answer aggregation and communication channels between workers (discussed later).

3.2. **Expertise of the Crowd**

Knowing (or not knowing) who composes the crowd is is only one dimension of selecting a crowd behind a system. Expertise can vary widely between crowd workers, as in any work settings. The source of the crowd often only controls how much we know in advance, rather than the range of skills that may be seen in the group.

Expert crowds are composed of people experienced in one or more relevant domains. Online platforms for these types of crowds are increasingly commonplace. For example, platforms such as oDesk (www.odesk.com), Elance (www.elance.com), Guru (www.guru.com), or Freelancer (www.freelance make it possible to hire easily professional contractors who specialize in anything from programming to design, project management, or advertising, among other domains. The level of expertise such workers bring may varying from a few months to several years. Such workers generally charge rates proportional to experience and ability.

General crowds are crowds that are not assumed to have any particular expertise related to the task they will be completing. While novice crowds lack specific, known expertise, they are much easier to find, recruit, and hire than experts. The cost is also typically proportionally lower.

3.3. **Design Implications**

Each different expertise level and source of crowds has a different effect on how the system must coordinate workers' efforts to be successful. Each of these considerations constitutes an axis of a continuous space, and as such can be mixed in various ways, contingent on the resources and platforms available to requesters. The space is roughly

Participant crowds consist of groups of people who are all involved in a specific endeavor. These may mix expert workers with non-experts – as is the case in Wikipedia, or FoldIt (Cooper et al., 2010) and eteRNA (Lee et al., 2014).

Open crowds provide a means of easily accessing the largest possible groups of workers to complete a task, but also introduce the possibility of spammers, workers trying to game the system, and even malicious workers. There is a similar trade-off in terms of expertise: lower expertise requirements leads to the ability to more quickly recruit a larger number of workers to a task (because there is a simpler-to-meet restriction imposed), but each worker might not be able to complete work as well as an expert.

If always-available on-demand access to a crowd-powered system is the goal, then moving more towards lower-expertise, unknown crowds opens the largest set of options (Figure 2). This might be why Mechanical Turk has become the primary resource for real-time systems. Another reason is the ease of access of the platform to these types of requests. Platforms that offer batch-processing services that a delayed by hours or days are clearly not viable options. While many services have properties that are not well suited to real-time work, and to our knowledge none specifically optimize for it, the availability and quality of special-built features for real-time systems is expected to improve as more successful systems arise.

Some of the key platform issues to consider are:
– Latency between a post and workers being able to accept (or be issued) a task.

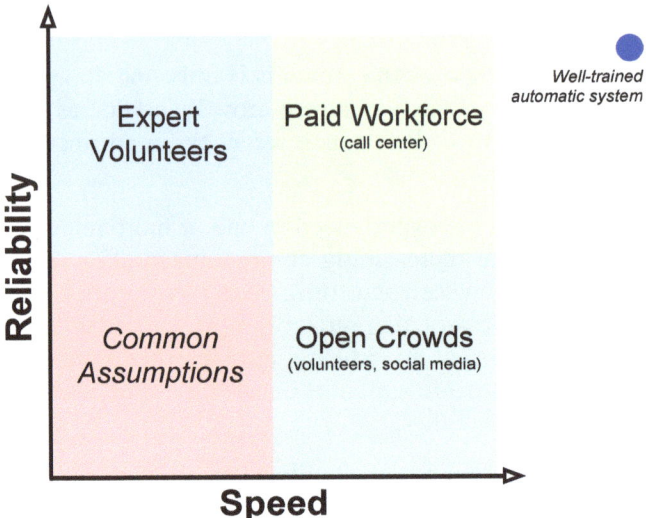

Figure 2. Speed and reliability trade-offs for different sources of crowds.

- The extent to which posts are collapsed into single entries.
- In the case that the platform uses posts that workers can select from: does the platform allow you to make workers aware of the fact that a task needs an immediate response?
- In the case that the platform assigns work directly to workers: does the platform allow you to notify the assignment process that the task needs an immediate response?

4.   DIVIDING TASKS

Once the workers are recruited, the first issue a system must address is how to assign tasks to them. Depending on the nature of the task, division into synchronous tasks may be appropriate, or coordinated turn-taking might be a better fit. In this section, we discuss the different division strategies available to system builders.

4.1.   Discrete Versus Continuous Tasks

Discrete real-time tasks are self-contained tasks that can be answered entirely by one worker – even if multiple workers may be assigned to increase answer quality or reduce speed. These tasks may include all required context, or may be part of an ordered sequence of tasks in which some context must be passed from one to the next (discussed in Section 6). Single-response tasks are the simplest instance of a discrete task, which require no additional context from prior responses.

Continuous real-time tasks are those that engage workers in a closed-loop interaction with the system. This, in turn, allows the system to engage other entities in its environment more fluidly. In many settings, this interaction is key: tasks such as when driving a car or a robot in a changing environment cannot reliably be done as a series of separate pieces, both because of the difficulty of maintaining situational context, and because it would make it hard to react to environmental factors such as pedestrians or falling rocks.

These tasks go outside of the traditional microtask model, but allow for just as much or even more flexibility for workers: they may join at any point, and continue working for as long as they wish. In most cases there will still be a minimum unit or work composing these tasks (e.g., typing a word) – this can be seen as the "microtask" within a continuous task.

However, as with other classes of crowdsourcing algorithms, simply dividing the task does not produce a more effective system – a means of combining these inputs is required (discussed in the next section). Accordingly, the selection of how to divide the task is intimately tied to the recombination strategy that can be used on the input.

4.2. **Stream Parallelism**

Stream parallelizable tasks (or *parallelized tasks*) are those that can be split into multiple sub-roles that can be completed by workers in parallel. These roles are carried out simultaneously over the same source-time period.

When workers arrive to a parallelized task with discrete sub-tasks, they are assigned a sub-task. This sub-task may be independent, as may be needed in general image labeling (Bigham et al., 2010; Russell et al., 2008), or part of a context-grouped or even ordered set.

When workers arrive to a parallelized *continuous* task, they are assigned a role and issued a task stream for that role. These roles may be defined statically by the system, e.g., in terms of a subset of controls in a larger interface, or dynamically by the workers themselves. For example, Legion:Gaming (Loparev et al., 2013) allowed each player to selectively control a portion of a video game controller.

The division of control within a task is not required. For instance, to let workers maintain complete situational context, Legion's control interface presented a single feedback stream (video) to all workers (Lasecki et al., 2011). As we will see, many tasks can be trivially stream-parallelized in this way in order to introduce redundancy.

4.3. **Temporal Division**

Temporally divisible tasks are those that can be effectively solved by dividing them into multiple subtasks that can be completed by different workers, where each subtask may depend on the output of earlier or later subtasks.

For instance, in order to caption live speech, workers would typically have to type upwards of 250 words per minute, or about 1250 characters per minute – something that exceeds the maximum recorded human typing speed (Wikipedia, 2014). Scribe (Lasecki et al., 2012) divides a streaming set of small ordered segments that are possible for people who can type quickly to caption. These tasks are also redundantly stream-parallelized, asking multiple workers to caption each segment as they arrive in order to produce a more reliable answer after recombination. This type of round-robin turn taking is an example of temporal division used to make the (computer-mediated) group performance better than that of any one constituent member. Another example can be seen in the real-time activity recognition and labeling system Legion:AR (Lasecki et al., 2013a).

Temporally dividing tasks in this way can lead to other benefits to workers' individual performance as well. For instance, TimeWarp (Lasecki et al., 2013b) introduced a collective process to Scribe that allows workers to each listen to audio streams at a reduced playback speed to increase precision and recall, while still decreasing latency.

Generally, the temporal division of tasks is useful when a task can be effectively completed by a worker for short periods of time, but not sustained for extended periods. Turns can also be determined based on semantic roles, with workers trading input control based on their knowledge or role. This can go beyond simple hand-offs and help to avoid context switching costs that arise when a single individual must accomplish multiple orthogonal tasks in order to make the system function as intended.

4.4.   Correction

Correctable tasks are those for which it may not be feasible for human workers to complete in real time, but in which errors can quickly be detected and fixed. Correction is typically a sequential (i.e., non-parallelizable) post-task that is run after a user or machine has generated an initial answer. It can be applied to almost any generation schema, but adds additional latency. This latency is based on the granularity of the task division – larger input segments will increase latency because workers must wait to get access to them. Correctable segments need not be the same size as initial-task segments, e.g., they may be smaller to reduce latency, or larger to provide more context.

As an example of a correction workflow, we can imagine an alternative approach to real-time captioning where one (or more) worker types the initial content, while other workers apply the corrections to the resulting caption stream. Even when content generation is handled by an automated process, the crowd can fill this role, such as in correcting captions generated by automatic speech recognition (Harrington and Vanderheiden, 2013). This example also shows that the correction approach can be used in conjunction with other task division approaches (e.g., stream-parallel or temporally divisible tasks) in order to make even the captioning problem feasible for non-experts.

4.5.   Combining Approaches

For each of the approaches above, most tasks can use a combination of at least one from each category, assuming the problem is well-suited to each approach individually. There are challenges involved in mixing approaches of course: time scales (segment lengths) may differ between different types of tasks, making them potentially hard to reconcile with one another, and the context needed for different divisions may vary.

There is also the issue of cost. Adding additional workers to any of these roles increases the overall cost of running the system, but adding a new process to a workflow is often a larger increase. For example, going from 4 to 5 workers in the round-robin process in Scribe is much more affordable than parallelizing the task between two workers in all four segment sets.

4.6.   Task Routing and Mediated Workflows

Regardless of the exact structure of the chosen division, intelligently routing workers can potentially increase the performance of each worker by matching them to a task that they can excel at. This can be done optimally, up to the limits of the information available about the workers' skill sets and task properties. Workers' skills are matched to the task requirements and the configuration yielding the best expected outcome is selected (Jung and Lease, 2013).

Computationally mediated workflows offer the promise of handling much of the coordination work that managers have in classical group-work settings (Retelny et al., 2014). Task routing can efficiently assign tasks to workers who are qualified for, hand-offs can be checked automatically and forwarded with the appropriate context, and new workflows can be generated dynamically based on prior experience. Decision theory can also be used to determine the phases of a workflow. If more workers are needed for a task, they can be routed there until the system is confident in an answer, and the workflow re-adapts accordingly (Dai and Weld, 2011; Lin et al., 2012).

Workflows can also be derived from existing structures of automated computation. For example, CrowdForge (Kittur et al., 2011) uses a Map-Reduce model to divide and recombine subtasks. The crowd might also be able to mediate its own workflows, using systems like TurkoMatic (Kulkarni et al., 2012), which recursively divides tasks based on worker input. The challenge for real-time systems is creating methods of letting workers hand off responsibility and delegate work in a very short amount of time.

4.7. **Modeling Routing and Decisions**

Ho, et al. study the *online task assignment problem*: given a collection of subtasks, they assume that each worker has a skill level for solving each task and a benefit level to the task requester (Ho et al., 2013; Ho and Vaughan, 2012). They present an online algorithm for assigning workers to tasks as they arrive and show the algorithm is, under certain assumptions, close to the optimal offline assignment. Karger also looks at this problem, but from a broader perspective that also considers how to best aggregate answers from competing workers (discussed in Section 5) (Karger et al., 2011). Heidari and Kearns study the problem of using the crowd itself to route discrete subtasks, where each worker, when presented with a subtask, may choose to perform it or pass it to another worker (Heidari and Kearns, 2013).

5. **ANSWER AGGREGATION**

Dividing tasks to help groups of people to provide more complete, accurate, and reliable responses, leads to the need for answer aggregation. Without answer aggregation, users receive multiple, potentially contradictory responses. While some systems, such as VizWiz, simply forward all responses back to a user, this leaves some uncertainty as to the reliability of the responses, especially when answers arrive one by one and could each potentially both be true. In this section, we discuss how answer aggregation algorithms, such as the consensus approaches for batch tasks presented in Section 2, can be applied to real-time systems.

The way the task has been divided determines much of how the aggregation process is structured. However, the aggregation algorithm used can be optimized for specific properties, such as reliability, speed, cost, and consistency.

5.1. **Optimizing for Reliable Responses**

The most common approach to ensuring reliable answers in crowd-powered systems is to increase the level of answer redundancy. This relies on the *wisdom of crowds*, which says that a group of people generally tend collectively towards the correct answer (e.g., after averaging (Surowiecki,

2005)), subject to the bias of the crowd towards a given task. The fine-grain distinction as to when and how people can be biased is beyond the scope of this paper.

However, the level of agreement alone is not the only available means of increasing answer reliability. In many cases, such as when soliciting open-ended responses, agreement between two different workers given a large answer space is a strong signal of correctness. This concept was used with synchronous workers for discrete tasks in The ESP Game (von Ahn and Dabbish, 2004), where workers each guessed a label for an image and the answer was accepted when both workers answered with the same label.

For continuous real-time systems, determining the quality of responses on the fly can be a challenge (Lasecki and Bigham, 2012b; Mashhadi and Capra, 2011). Assigning additional workers to a workflow in order to get redundant responses can have negative impacts on the system. If a worker set that is just large enough to ensure high confidence is chosen in advance to complete tasks in parallel, then the work can be completed as fast as the slowest response. However, it is hard to know just how large the worker set needs to be and, often, reliable agreement with fewer workers than the pre-determined level is possible. On the other hand, if an iterative approach is used to terminate execution as soon as a high-reliability answer is found, it is likely to add undue latency.

Trading off between reliability and speed has been explored in previous work (Singh et al., 2012; Horton and Chilton, 2010), but remains an open problem. Incentives based on early, later-confirmed responses have been one effective means of incentivizing workers in real-time systems to optimize this tradeoff (Lasecki et al., 2013; Singh et al., 2012). Algorithmic optimization approaches to balancing quality under time constraints have also been developed (Boutsis and Kalogeraki, 2013).

5.2. Optimizing for Rapid Responses

If a quick answer is the most important aspect of a system, first-response approaches are usually the most effective. VizWiz provided the first result available to users, relying on the fact that end users can typically identify incorrect answers easily for verification (e.g., the answer "A can" to the question "How many calories are in this package of food?" is incorrect). Another contributing factor is that, anecdotally, workers provide more reliable answers when presented with tasks that do social good. More generally, larger sets of workers benefit from natural variance yielding at least one faster worker in the typical case than any single worker making repeated contributions.

5.3. Optimizing for Affordable Answers

Cost is a critical factor in making a deployed system sustainable. To minimize the cost, the total worker time used must be reduced, which often means that the total number of task touches needs to be decreased, even if as a consequence each worker must complete a slightly longer task. Doing so can reduce the context switching overhead that workers incur when joining a new task.

To minimize the number of tasks that need to be issued by the system in settings where providing even occasional inaccurate responses to users can be detrimental effects, measures of response confidence can be used. Response sets can vary widely between tasks. For simple tasks with clear correct answers, nearly all workers may be in agreement from the beginning. On the other hand, edge cases, or otherwise difficult or ambiguous tasks might yield conflicting sets of responses with no clear winner, meaning only a response from a larger crowd might be considered reliable. Since this is sometimes unavoidable in collective intelligence systems, our goal becomes to answer each question in as few responses as possible with a given level of confidence.

Iterative workflows offer an effective structure for approaching this problem: by evaluating worker agreement at each iterative step, we can cut off processing when a confidence threshold is reached (Karger et al., 2011). This threshold can have based on standard statistical confidence measures, such as p value or confidence interval. However, as mentioned above, iterative workflows can be too slow for many real-time tasks. As such, they offer a solution only in cases where the delay incurred is acceptable, or where the initial response provided to a user can be updated as new answers arrive (e.g., data visualization or captioning).

5.4. Optimizing for Consistent Responses

Optimizing for quick, reliable, or affordable responses can enable systems that respond helpfully to users. But not all problems require only a single answer. Thus far, we've focused on performance criteria that are largely assuming that answers each exist in isolation. In practice, some tasks require the system to maintain state or understand the implications of prior actions (e.g., when creating a virtual agent (Rossen and Lok, 2012; Borish et al., 2014; Lasecki et al., 2013). This idea introduces a new measure of answer quality: *consistency*.

Consistency can be defined as producing responses that do not conflict with prior responses. This definition heavily depends on the task at hand. For conversational interaction for instance, responses should not disagree with prior claims about the system's beliefs, or repeat parts of a conversation that a single person would not (for instance, saying "hello" again after initial introductions have already been completed).

Legion explored a number of approaches to real-time interface control, and found that one of the most reliably fast and accurate was selecting a representative leader from the crowd once each small time unit (Lasecki et al., 2011). By doing this, responses best represent the will of the collective, while avoiding self-contradicting actions from turn to turn.

More generally answer consistency be maintained by ensuring the crowd is aware of prior context, and have the tools (usually communication paths) to avoid conflicting or repeating the answers generated by other workers in settings, especially when their answer will not be aggregated automatically. We discuss both context maintenance and communication channels later in this paper.

5.5. Formal and Theoretical Approaches

The main theoretical framework for studying information aggregation among heterogeneous agents is social choice theory. This framework generally consists of two collections: one of agents (i.e., the workers in a crowdsourced setting) and one of *alternatives*. Each agent is assumed to have a (partial) ranking of the alternatives. Often the agent must also report these rankings to an aggregation system. This can often be an arduous task, because the number of alternatives can be very large. Social choice theory typically addresses various ways in which the preferences of individual users can be aggregated into a single preference list. Such methods are called *social welfare functions* or, when they return merely the top choice, *social choice functions*. Typical questions on social choice functions involve whether the function satisfies certain fairness criteria. Famously, Arrow showed that no social choice function must fail at least one reasonable fairness criterion (Arrow, 2012).

Typically, social choice theory considers only one election at a time and assumes that all agents vote simultaneously, as is the case in standard political elections. Recently, Parkes and Procaccia use Markov decision processes to extend elections into multiple rounds in which the preferences of the individuals represent the state of the system at any given time, and each election changes the preferences of each individual (Parkes and Procaccia, 2013). (Hemaspaandra et al., 2012, 2014) consider elections in which agents vote *sequentially*, rather than in parallel.

Another approach to achieving consensus is to treat the task as a machine learning problem in which workers provide partial or noisy answers and the aggregation mechanism filters the best answer from the workers' input. For instance, Snow et al. study the accuracy of Mechanical Turk workers in text annotation tasks. They propose a method for improving accuracy using expectation maximization and a training set of data annotated by task experts (Snow et al., 2008). Kamar et al. used partially-observeable Markov decision processes (POMDPs) in *consensus tasks* (Kamar et al., 2012; Kamar and Horvitz, 2012), which they define as tasks in which multiple workers propose answers to a single question and the goal of the system is to come up with an optimal policy that either recruits a new worker or selects one of the answers already proposed, by considering the cost of adding more workers against the expected improvement in the quality of the answers. See also (Dalvi et al., 2013; Karger et al., 2011; Carpenter, 2011; Liu et al., 2012; Raykar et al., 2010; Sheng et al., 2008; Zhou et al., 2012)

6. MAINTAINING CONTEXT

When dividing and later aggregating tasks from a larger job, we need to ensure that workers have enough context to complete their task. Context might be critical, such as remembering previously traversed paths in a robot navigation setting (Lasecki et al., 2011, 2012) – or a tool to improve efficiency and user experience, such as maintaining a consistent tone and topic history over multiple interactions with a conversational assistant (Lasecki et al., 2013). For both of these cases, the context shared between the crowd might be *explicit* content that must be read by each worker, or *implicit* in the observations of prior events.

6.1. Explicit

The most direct means of maintaining context between multiple interactions is to show relevant information to later workers. This is conceptually simple, but presents a number of challenges: how to prevent this information from becoming overwhelming to process over many interactions? How do we determine what is most important or worth passing on? How can we prevent creating and consuming this information from adding latency to our task?

Note taking is a widely used tool for remembering key details and pieces of information for later. Many strategies are evolved on an individual basis for best tracking this information (Van Kleek et al., 2009). However, scaling this to a collective, where different people might be more adept at certain strategies, and there is no one single overseer who understands all aspects of the problem in order to judge what is most important.

Existing crowd-powered systems have seen successful approaches based around special purpose note-taking systems (Lasecki et al., 2013). Automated systems might also act to augment the crowd's ability to remember and process large sets of information over long periods of time.In a puzzle-solving setting, message passing between workers has also been observed as a means of preventing thrashing between states, leading to more efficient solutions (Zhang et al., 2012). If the system has enough information about the task, it may be able to explicitly direct workers to hand off certain information to future workers (even when the information cannot be extracted automatically), such as with expert crowds in flash teams, where links between modular tasks define deliverables for use by future teams, and parallel tasks require more open-ended coordination and message passing during the task execution (Retelny et al., 2014).

However, this is an understudied problem. Few crowd-powered systems exist that meaningfully support the same user over multiple sessions. The most important elements to consider are:

- **What is important?** Predicting if a fact will be important is within the capability of workers in many (but not all) situations. Forecasting future events is even an active area of research in crowdsourcing (Forlines et al., 2014).
- **Who will record the information?** Crowd-curation can involve a separate task and workflow, but can yield better results than automated summaries. By combining human and automated approaches, such as having people identify what is important and having the machine summarize textual information, the cognitive effort and time required for human workers can be reduced.
- **How much can a worker process quickly?** This varies significantly per-task, but it is important to consider the latency impact of having to process and understand the information captured about prior sessions. Often, this load can be distributed over time, or even included as part of the existing tutorial phases to reduce the direct overhead.
- **How will the task be structured?** Depending on the task and type of information that needs to be recorded, the same workforce might be used to complete the core task and record information, or two separate groups might be used. Separating the task makes the problem of increasing latency and worker cognitive load more manageable, but increases operating costs.

6.2. Implicit

Implicit context might be shared by direct observation of other workers, or the results of their actions. It relies heavily on workers being willing to observe and assess their surroundings for clues to prior events that have an impact on their potential responses. While this may seem impractical for microtask workers who must complete tasks at a quick pace to earn a livable wage, prior work has observed workers waiting to see what others do before proceeding to propose their own answers in the presence of others (Lasecki et al., 2012). Crowd workers also remember some task details between different sessions spanning hours or days, so providing incentives for returning can afford some level of consistency between sessions (Lasecki et al., 2012; Bigham and Lasecki, 2014).

Explicit context maintenance is easier to design for than implicit mechanisms in most cases, because the tools used usually have more predictable uses and outcomes. However, this comes at the cost of increased latency to interact with the mechanism, or increased cost for additional workers to complete tasks in parallel. Thus, if implicit cues can be used, they can frequently provide a much smoother and less costly means of transferring information.

7.   COMMUNICATION CHANNELS

Crowd-curated context can be seen as a case of one-way (potentially mediated) worker-to-worker communication with large time delays. As discussed in the previous section, communication can be very helpful for communicating context and other task needs to later workers. In this section, we explore how communication during a task can help tasks get completed quicker, and more reliably.

Communication in collaborative settings has been well studied in management theory, organizational behavior and psychology, human-computer interaction, and other fields. However, within crowdsourcing, the collaborative nature of tasks is often overlooked, perhaps because of a combination of the lack of native platform support (on almost any existing crowd marketplace) and the potential for collusion between workers to inhibit the wisdom-of-crowds effect.

7.1.   Design Implications of Communication

Allowing workers to communicate with one another can be beneficial in multiple ways. Since workers are generally well-intentioned, but sometimes do not understand specific task rules, guidelines, or objective, other workers can help increase adherence to these rules and actually improve the performance of workers who might otherwise contributed noise (Lasecki et al., 2013). Extending this idea to a formal feedback framework has been shown to be effective at helping workers improve over time (Dow et al., 2012) – an important consideration when systems will be deployed and rely on a perpetual crowd workforce.

Communication can also allow workers to build on one another's work so far, and prevents conflicts from arising between workers. This can be of benefit in settings such as the collaborative search process in Chorus: workers could each claim a specific portion of a task, and then focus on that sub-task without worrying that another worker might complete it before them and nullify their contribution. The tradeoff here is that this could reduce answer speed and diversity if only one user searched for an answer, but reduces the potential for uncertain payoff and redundant responses to some aspects of a task while starving others. In an ideal case, communication can be used to effectively parallel parts of the task, while helping workers extend each other's efforts.

However, allowing for direct communication also has drawbacks. Workers might collude with one another to increase their odds of payment when majority decisions are used, or more vocal workers might persuade others to agree with their answer without regard to its correctness. An additional problem is well known to online platforms of all types: spammers.

Paralleling our discussion of context, we will look at creating both explicit and implicit channels for worker communication in this section.

7.2.   Explicit

Creating explicit communication channels can range from text to audio or video, though the latter is far more rare because it's harder to host and mediate. This leaves text as the dominant form of direct communication behind most any crowd-powered system. By creating an explicit channel for communication, the worker experience can be more carefully designed, and workers will have a known resource to look to, rather than having to go without support or miss discovering an implicit communication channel.

To address the drawbacks stemming from direct communication, workers can be split into sub-groups. For tasks that would otherwise work without communication, this provides workers with a support group, while not biasing the entire crowd by any one individual. Similarly, it provides for speed and diversity in sub-tasks by mixing workers from different groups, while also ensuring that starvation is minimized since workers within a sub-group will be able to avoid conflict. To prevent spamming within sub-groups, rate limiting or feedback from other users can be used to deter over-posting.

In settings where crowds large enough to be divided into effective sub-groups are not practical, other mediation strategies can be used to prevent collusion and bias. For example, random spot-checking responses can provide confirmation that workers are giving helpful, differentiated answers. Spot-checking uses the crowd to act as a force-multiplier for the user, rather than fully-autonomous control (discussed more in the next section). In some cases, tasks with known gold-standard answers can also be inserted into a workflow to detect when workers might be cooperating to the detriment of quality. If this is observed, then communication might need to be reduced between workers in favor of more indirect means.

Not all channels of direct communication need be intentional. Chorus observed workers effectively using 'proposed' messages (which are only visible to other workers) to chat amongst one another without forwarding the response to users by voting on them. This was used for good (e.g., helping other workers understand the task) in some cases, and to the determent of the system in others (e.g., spammers). Interestingly though, this method of communication required workers to coordinate and trust one another enough to not forward the internal messages to users, and to actually pay a small cost (in terms of lost bonus) to propose an answer they knew would not be voted through. This suggests that potential mechanisms for incentivizing workers to contribute only the most important comments to one another may be effective.

7.3. Implicit

Implicit communication channels do not allow direct worker-to-worker contact, but instead display some feedback to workers about what others are doing. Implicit communication channels can frequently be overlooked when designing a system, or even might be impossible to remove. For example, Legion explored a visual feedback mechanism in a robot control task that let workers know what others in the crowd had done in the previous time-step if it differed from theirs. However, this quickly resulted in workers all trying to follow the crowd's previous decision, leading to many experimental trials ending in a longer series of repeated actions (such as spinning in circles).

On the other hand, using only the observation of the outcome of the task (instead of the exact input), which is seen by workers in the movements of the robot in the video and is essentially impossible to completely remove, led to a learned group behavior that let the crowd collective control the robot for over an hour without forgetting the intended path, despite the high level of turnover in the crowd and the lack of instructions to anyone but the first small set of workers who had begun the task.

Indirect cues such as showing the crowd's final response are not without their use however. Showing aggregate information about the crowd's collective decision also provides a means of allowing more adept workers a chance to compensate for the actions of inattentive or lazy workers (Lasecki and Bigham, 2012a). This also requires an attention to the incentives provided to workers. Even more subtle, indirect forms of feedback include displaying results from workers (individual or aggregate) to other workers (Lasecki et al., 2011, 2013, 2014).

7.4.   Theoretical Aspects of Communication

Communication problems can be roughly divided into two categories: collusion and coordination. Collusion in the form of voter *manipulation* has been widely studied in social choice theory. See (Faliszewski et al., 2010) for a discussion of this and related topics.

Coordination is when the workers collaborate on a shared goal and individually have incomplete information about the situation, but can benefit from sharing information. We can characterize such problems in terms of Markov decision problems and their extensions to multiagent settings. When each worker is aware of all the observations and actions available to all the other workers, the variant is called a *multiagent POMDP* (MPOMDP), otherwise, it is a *decentralized POMDP* (Dec-POMDP). In both cases, workers are assumed to share a single reward and thus benefit from some communication. To model the case where each worker has a distinct reward, partially-observable stochastic games (POSGs) are used. While workers in practice almost always have individual incentives for participating, when studying the benefits of communication from a theoretical perspective, it is common to focus on collective benefits. We discuss the problem of structuring incentives to get desired outcomes from the workers with selfish motives in section 8.

The basic question asked of POMDP models is: what are the optimal or equilibrium policies? Determining an optimal policy in this case is NEXP-hard for Dec-POMDPs. To address this computational complexity barrier, the usual approach is to adopt the same approach taken to heuristically solve POMPDPs. Often, to improve the tractability of the problem, additional restrictions are assumed, say, in terms of *coordination graphs* (Proper and Tadepalli, 2009) or branch-and-bound heuristics (Spaan and Oliehoek, 2012). See also (Messias et al., 2011; Goldman and Zilberstein, 2004; Boutilier, 1999).

8.   REWARDING WORKER CONTRIBUTIONS

The ultimate goal of a reward/penalty scheme is to provide workers with incentives that guarantee a desired level of effort, quality, or task completion time. However, in practice it is often hard to evaluate the quality of the work or to know what truly motivates workers. If there was a right answer to compare to, then there would be no need for worker input.

A lot of work has been done in the area of studying how to incentivize work in different scenarios. The literature in crowdsourcing has also followed suit. However, many of the most effective incentive mechanisms rely on task configurations that are either too difficult or require too much time to understand, or are not compatible with real-time systems in their original implementation. We discuss strategies that can be used in real-time to reward helpful work.

8.1.   Payment Limitations

One issue that reduces flexibility in reward mechanisms is the limitation of the crowd platform itself. For example, Mechanical Turk (as well as most others) does not provide a means of marking a task as "completed in good faith, but incorrect". This means that workers who try and fail to get a correct answer can only be either rejected or paid the full amount of the task without feedback. This leads to a culture where requesters generally just pay for completed tasks regardless of quality, and rejections are seen as an exceptionally strong response to incorrect work.

From a payment side, this can be partially mitigated using bonuses, if supported by the platform – a base payment for taking the task and genuinely trying to provide a valid answer, and a bonus can be paid for higher quality work. This also allows for much more flexibility in the incentive mechanisms that can be used. In this section, we will assume that the platform being used allows for delayed rewards.

8.2.   Design Implications of Reward Strategies

As we discuss reward strategies (incentive mechanisms) it is important to keep in mind that these payments are not just a means of motivating workers to do good work – they often do that even in the face of poorly designed rewards – but also to provide workers fair pay for the effort they have put into the task. The key is knowing when this effort has really been put in, and how much to pay.

"Correctness" is typically determined in one of two ways: either by using agreement between a sufficient number of workers, or using gold-standard tasks to periodically inspect worker responses. While these methods often include some delays in order to determine the right answer, heuristics can be used to provide approximate immediate feedback to workers that they are on the right path, even if the result does not turn out to be perfect. In Scribe, automatic pre-checks (i.e., spell checking) and agreement are used to give immediate feedback as to whether the system thinks the user is correct in typing a word, even before the system is really confident in the final answer. Delayed rewards (bonuses) can then be used to compensate for the cases when the heuristic fails to reward a worker for helpful input immediately.

To maintain speed, microtasks (as with any fixed-value work) push workers to balance between speed and efficiency – spending more time on a task increases the chances that it will be considered well-done, while spending less time increases the hourly rate of the task and allows for higher overall income. To maximize economic value, the worker's goal then becomes to do the task as quickly as it can be done correctly. The problem lies in the worker's ability to estimate all of the factors in this balance, which is difficult even if all of the information required has been made available to the worker.

HiveMind (Singh et al., 2012) is an incentive mechanism that used this basic balance to vary the rate at which answers are elicited by rewarding early answers relative to later ones by a larger or smaller factor, and rewarding answers more than votes of agreement. Larger payment for early answers relative to late answers leads workers to contribute fewer answers, and agree with existing answers more often. More even payments over time lead workers to continue to contribute new answers for the chance to earn a higher pay. The requester can thus adjust the rewards depending on what is of higher value to the system.

Another frequent, and perhaps more fundamental issue is that the reward mechanisms being used are not clear to workers, meaning that workers cannot correctly account for them when considering what actions to take. To make them clear, strategies with direct mappings to visible actions should be used. If behind-the-scenes information has to be used to compute the reward (e.g., basing rewards on the input of other workers), some indication of this should be shown. Explanations of the process used, as well as tutorials that allow workers to experiment with how rewards are given in practice have also proven to be effective ways to help workers understand the task and its rewards better (Lasecki et al., 2013,b, 2014).

8.3. **Joint Rewards**

Joint rewards are needed when the collective behavior cannot just be the sum of individually optimized behaviors. For example, encouraging workers to give the correct answers to a navigation problem can be done using individual rewards (because we assume we can compare to other workers), but using self-correction (Lasecki and Bigham, 2012a) requires a joint success reward to see if workers were able to collectively able to correct for the group action (discussed in Section 7).

Another example of joint rewards is meta-rewarding. For instance, Legion used a global time limit with a reward multiplier for faster times. By creating a two-tier optimization, workers can be encouraged to agree and rewarding them for the time they spent on the task, while also rewarding the group for finishing faster.

More generally, rewards help to provide workers with not only a wage, but a means of getting feedback on their performance (even if that feedback is often very noisy, such as on Mechanical Turk). There are a number of non-monetary ways to provide feedback in real-time settings, many of which fall under the umbrella of gamification.

8.4. **Theoretical Models**

The most common theory for reasoning about such problems is *mechanism design*. Game theory asks what the best strategy is for an individual to take when the outcome depends on how the other individuals (who themselves are acting strategically) play, whereas mechanism design asks how to best design the game so that the workers' optimal strategies lead to the outcomes that the designer desires. It is most commonly used in the context of auctions, and there is a substantial body of work that looks at crowdsourcing from this perspective (DiPalantino and Vojnovic, 2009; Boudreau et al., 2011; Moldovanu and Sela, 2001, 2006; Archak and Sundararajan, 2009; Chawla et al., 2012). However, this work does not focus on real-time settings, and generally assumes that the response timescale is much longer and that finding high-quality solutions is a matter of time and effort. Also, such settings typically involve a single round. In real-time settings, workers have little time to respond, meaning one must regard the knowledge and beliefs of the workers as relatively static. A real-time task often involves multiple rounds.

Models in which the actions, observations, and rewards of multiple actors are explicitly defined are called partially observable stochastic games (POSGs) (Kuhn, 1953). In POSGs there may be no well-defined globally optimal (i.e., dominant) joint strategy. Instead, strategic equilibria must be found, but it can be NEXP-hard (Bernstein et al., 2002) to compute these, even offline and with complete knowledge of the model. TurKontrol (Dai and Weld, 2011) used a POMDP model in order to optimize iterative task completion online.

9. **ADDING AUTOMATION**

Thus far, we have discussed how to create systems that use computer-mediated groups of people (crowds) to go beyond what automated systems can do by themselves. However, beyond mediating interactions between workers and routing tasks, there are many automated solutions that target these problems. Their solutions may not work in a wide enough variety of settings to create the desired reliability, or may only give partial solutions, but they can be integrated into crowd-powered systems as a means of making human workers' tasks easier.

9.1. Design Implications of Automated Assistance

Crowdsourcing can be seen as a *scaffold* for intelligent systems: a way to support systems in situations where they cannot operate independently. This provides users with a seamless experience, while providing a means of training the automated system to be more capable over time (learning by demonstration (Osentoski, Crick, Jay, and Jenkins, Osentoski et al.; Lasecki et al., 2011).

Computers can also go beyond the limitations of people. While people are great sources of general-purpose processing – able to understand context and disambiguate situations using many different facets of a problem – they are not nearly as fast as computers at well-defined data-processing tasks. Using computers to contribute information, such as search results or patterns found in a dataset, allows groups containing both human and machine contributors to outperform either individually. These *hybrid* systems leverage the fact that the collaboration methods used in crowdsourcing are often flexible enough to handle automated systems as just another set of workers, with a different set of mistakes and biases.

Automating systems over time also provide a compelling argument for the scalability of crowd-powered systems: people are sources of "live" training data for future fully-automated systems. With this in mind, the input mechanisms for and workers interfaces can be designed in such a way that they elicit information that not only answers the question posed, but also provides a formal enough notion of the information for computers to parse. For example, ARchitect (Lasecki et al., 2014) extended Legion:AR (Lasecki et al., 2013a) with an interface that let workers convey the dependency structure of a set of actions by simply answering a set of yes-no questions. This allows the system to generalize observed and labeled activities the crowd produces and helps the system to learn accurate patterns with fewer examples.

9.2. Roles of Automation

To help provide a framework for integrating automated systems into crowd-powered systems, we highlight the following potential roles that automation can commonly fill:

9.2.1. Support

One of the easiest means of integrating automated input into a system's workflow is to use AI as a "first pass". By producing an initial guess automatically, many problems can potentially be solved immediately, and at most use worker input to confirm their accuracy. This reduces the time and effort that workers must contribute, thus making the system cheaper to run. However, editing large responses can itself become more challenging than simply creating a new response if the error rate is high enough. As such, when no algorithm is available that can achieve high accuracy consistently, or has a reliable confidence metric for selecting the cases where error is expected to be low, then it might be more effective to abstain from providing this input altogther.

9.2.2. Augment

Even when approaches are not available to wholly solve any instances of the problem facing the system, automated systems can be made available as support tools. Web search tools can help workers access information, even before they explicitly request it, and prediction systems can make response suggestions based on other workers' input. Even simple tools such as spellcheck can improve the quality and speed of responses significantly.

9.2.3. Guide

Beyond directly helping workers with completing their task, automated systems can be used to highlight portions of the task that need the most attention. This might be based on features within the task instance itself, based on other workers' input to the same task, or based on aggregate information about previously observed work. This type of guidance could also be provided by a human worker, but the cost of giving a worker context and the higher response latency of a human contributor typically makes an automated approach a more viable choice. Note the distinction between support via in-task guidance, and task-level routing (discussed in Section 4) where tasks are assigned to workers based on system needs and, if known, the workers' skill sets.

9.2.4. Complete Tasks

Automated systems can also be used to entirely solve some tasks without involving a human at all. This use-case is tightly tied with the task routing method being used, which must have a means of assessing the capability of the automated system relative to the task it is faced with. If the systems' confidence is high enough that the problem can be solved automatically, then the task is not routed to a human worker. When it is not, the system employs an active learning approach to get answers from workers. We include this role here because it relies on the idea that automated systems can fill equivalent roles as people and have work selectively routed to them when they are capable. The capabilities of each AI can vary (just as with people), and the routing algorithms used at the hybrid system level should be able to remain almost entirely unchanged.

New methods for combining these roles (and more) continue to be created. The exact ideal mix depends on the task at hand, the capabilities of current automated approaches, and the cost trade-off between using human input and developing a custom automated solution.

10. CONCLUSION

In this paper, we have discussed an architecture for crowd-powered intelligent systems that can respond to users in real time. For each component in the architecture, we outline key approaches from relevant literature and anecdotal experience, design implications and trade-offs for each approach, and examples of systems that have effectively used the approach. We then ground our discussion in the models underpinning these components in order to provide more context for future work to build upon.

As with any design space, a complete set of options cannot be enumerated. Instead, our goal is to inform system builders who wish to augment interactive systems with human intelligence from the crowd about how to design systems in this complex space that is only beginning to be explored.

11. ACKNOWLEDGEMENTS

This work has been supported by the National Science Foundation (under awards #IIS-1149709, #IIS-1218209, and #SES-1111016), Google, Yahoo, a Sloan Foundation Fellowship, and a Microsoft Research Ph.D. Fellowship.

The authors would like to thank all those who have contributed to this line work both via tool creation and insightful discussion, including Christopher Miller, Mitchell Gordon, Donato Borello, Rachel Wesley, Ece Kamar, Matthew Murphy, and others.

12. REFERENCES

Archak, N and Sundararajan, A. (2009). Optimal Design of Crowdsourcing Contests.. In *ICIS*. 200.

Arrow, K. J. (2012). *Social choice and individual values*. Vol. 12. Yale university press.

Bernstein, D. S, Givan, R, Immerman, N, and Zilberstein, S. (2002). The complexity of decentralized control of Markov decision processes. *Mathematics of operations research* 27, 4 (2002), 819–840.

Bernstein, M. S, Brandt, J, Miller, R. C, and Karger, D. R. (2011). Crowds in Two Seconds: Enabling Realtime Crowd-powered Interfaces. In *Proceedings of the 24th Annual ACM Symposium on User Interface Software and Technology (UIST '11)*. ACM, New York, NY, USA, 33–42. DOI:http://dx.doi.org/10.1145/2047196.2047201

Bernstein, M. S, Karger, D. R, Miller, R. C, and Brandt, J. (2012). Analytic Methods for Optimizing Realtime Crowdsourcing. *CoRR* abs/1204.2995 (2012).

Bernstein, M. S, Little, G, Miller, R. C, Hartmann, B, Ackerman, M. S, Karger, D. R, Crowell, D, and Panovich, K. (2010). Soylent: A Word Processor with a Crowd Inside. In *Proceedings of the 23Nd Annual ACM Symposium on User Interface Software and Technology (UIST '10)*. ACM, New York, NY, USA, 313–322. DOI:http://dx.doi.org/10.1145/1866029.1866078

Bigham, J. P, Jayant, C, Ji, H, Little, G, Miller, A, Miller, R. C, Miller, R, Tatarowicz, A, White, B, White, S, and Yeh, T. (2010). VizWiz: Nearly Real-time Answers to Visual Questions. In *Proceedings of the 23Nd Annual ACM Symposium on User Interface Software and Technology (UIST '10)*. ACM, New York, NY, USA, 333–342. DOI:http://dx.doi.org/10.1145/1866029.1866080

Bigham, J. P and Ladner, R. E. (2011). What the Disability Community Can Teach Us About Interactive Crowdsourcing. *interactions* 18, 4 (July 2011), 78–81. DOI:http://dx.doi.org/10.1145/1978822.1978838

Bigham, J. P and Lasecki, W. S. (2014). Crowd Storage: Storing Information on Existing Memories. In *Proceedings of the SIGCHI Conference on Human Factors in Computing Systems (CHI '14)*. ACM, New York, NY, USA, 601–604. DOI:http://dx.doi.org/10.1145/2556288.2557159

Borish, M, Cordar, A, Foster, A, Kim, T, Murphy, J, and Lok, B. (2014). Utilizing Real-time Human-Assisted Virtual Humans to Increase Real-world Interaction Empathy. In *Kansei Engineering & Emotion Research (KEER '14)*. 15.

Boudreau, K. J, Lacetera, N, and Lakhani, K. R. (2011). Incentives and problem uncertainty in innovation contests: An empirical analysis. *Management Science* 57, 5 (2011), 843–863.

Boutilier, C. (1999). Sequential optimality and coordination in multiagent systems. In *IJCAI*, Vol. 99. 478–485.

Boutsis, I and Kalogeraki, V. (2013). Crowdsourcing under Real-Time Constraints. In *Parallel Distributed Processing (IPDPS), 2013 IEEE 27th International Symposium on*. 753–764. DOI:http://dx.doi.org/10.1109/IPDPS.2013.84

Callison-Burch, C. (2009). Fast, Cheap, and Creative: Evaluating Translation Quality Using Amazon's Mechanical Turk. In *Proceedings of the 2009 Conference on Empirical Methods in Natural Language Processing: Volume 1 - Volume 1 (EMNLP '09)*. Association for Computational Linguistics, Stroudsburg, PA, USA, 286–295.

Carpenter, B. (2011). A Hierarchical Bayesian Model of Crowdsourced Relevance Coding.. In *TREC*.

Chawla, S, Hartline, J. D, and Sivan, B. (2012). Optimal crowdsourcing contests. In *SODA*. 856–868.

Chilton, L. B, Sims, C. T, Goldman, M, Little, G, and Miller, R. C. (2009). Seaweed: A Web Application for Designing Economic Games. In *Proceedings of the ACM SIGKDD Workshop on Human Computation (HCOMP '09)*. ACM, New York, NY, USA, 34–35. DOI:http://dx.doi.org/10.1145/1600150.1600162

Cooper, S, Khatib, F, Treuille, A, Barbero, J, Lee, J, Beenen, M, Leaver-Fay, A, Baker, D, Popović, Z, and others, . (2010). Predicting protein structures with a multiplayer online game. *Nature* 466, 7307 (2010), 756–760.

Dai Peng, M and Weld, D. S. (2011). Artificial intelligence for artificial artificial intelligence. In *Twenty-Fifth AAAI Conference on Artificial Intelligence (AAAI 2011)*.

Dalvi, N, Dasgupta, A, Kumar, R, and Rastogi, V. (2013). Aggregating crowdsourced binary ratings. In *Proceedings of the 22nd international conference on World Wide Web*. International World Wide Web Conferences Steering Committee, 285–294.

DiPalantino, D and Vojnovic, M. (2009). Crowdsourcing and all-pay auctions. In *Proceedings of the 10th ACM conference on Electronic commerce*. ACM, 119–128.

Dow, S, Kulkarni, A, Klemmer, S, and Hartmann, B. (2012). Shepherding the Crowd Yields Better Work. In *Proceedings of the ACM 2012 Conference on Computer Supported Cooperative Work (CSCW '12)*. ACM, New York, NY, USA, 1013–1022. DOI: http://dx.doi.org/10.1145/2145204.2145355

Faliszewski, P, Hemaspaandra, E, and Hemaspaandra, L. A. (2010). Using complexity to protect elections. *Commun. ACM* 53, 11 (2010), 74–82.

Forlines, C, Miller, S, Guelcher, L, and Bruzzi, R. (2014). Crowdsourcing the Future: Predictions Made with a Social Network. In *Proceedings of the SIGCHI Conference on Human Factors in Computing Systems (CHI '14)*. ACM, New York, NY, USA, 3655–3664. DOI:http://dx.doi.org/10.1145/2556288.2556967

Goldman, C. V and Zilberstein, S. (2004). Decentralized control of cooperative systems: Categorization and complexity analysis. *J. Artif. Intell. Res.(JAIR)* 22 (2004), 143–174.

Harrington, R. P and Vanderheiden, G. C. (2013). Crowd Caption Correction (CCC). In *Proceedings of the 15th International ACM SIGACCESS Conference on Computers and Accessibility (ASSETS '13)*. ACM, New York, NY, USA, Article 45, 2 pages. DOI: http://dx.doi.org/10.1145/2513383.2513413

Heidari, H and Kearns, M. (2013). Depth-Workload Tradeoffs for Workforce Organization. In *First AAAI Conference on Human Computation and Crowdsourcing*.

Hemaspaandra, E, Hemaspaandra, L. A, and Rothe, J. (2012). Online Voter Control in Sequential Elections.. In *ECAI*. 396–401.

Hemaspaandra, E, Hemaspaandra, L. A, and Rothe, J. (2014). The complexity of online manipulation of sequential elections. *J. Comput. System Sci.* 80, 4 (2014), 697–710.

Ho, C.-J, Jabbari, S, and Vaughan, J. W. (2013). Adaptive task assignment for crowdsourced classification. In *Proceedings of the 30th International Conference on Machine Learning (ICML-13)*. 534–542.

Ho, C.-J and Vaughan, J. W. (2012). Online Task Assignment in Crowdsourcing Markets.. In *AAAI*.

Horton, J. J and Chilton, L. B. (2010). The Labor Economics of Paid Crowdsourcing. In *Proceedings of the 11th ACM Conference on Electronic Commerce (EC '10)*. ACM, New York, NY, USA, 209–218. DOI:http://dx.doi.org/10.1145/1807342.1807376

Ipeirotis, P. G. (2010). Analyzing the amazon mechanical turk marketplace. *XRDS: Crossroads, The ACM Magazine for Students* 17, 2 (2010), 16–21.

Jung, H. J and Lease, M. (2013). Crowdsourced Task Routing via Matrix Factorization. *CoRR* abs/1310.5142 (2013).

Kamar, E, Hacker, S, and Horvitz, E. (2012). Combining Human and Machine Intelligence in Large-scale Crowdsourcing. In *Proceedings of the 11th International Conference on Autonomous Agents and Multiagent Systems - Volume 1 (AAMAS '12)*. International Foundation for Autonomous Agents and Multiagent Systems, Richland, SC, 467–474.

Kamar, E and Horvitz, E. (2012). Incentives for truthful reporting in crowdsourcing. In *Proceedings of the 11th International Conference on Autonomous Agents and Multiagent Systems-Volume 3*. International Foundation for Autonomous Agents and Multiagent Systems, 1329–1330.

Karger, D. R, Oh, S, and Shah, D. (2011). Iterative learning for reliable crowdsourcing systems. In *Advances in neural information processing systems*. 1953–1961.

Kittur, A, Nickerson, J. V, Bernstein, M, Gerber, E, Shaw, A, Zimmerman, J, Lease, M, and Horton, J. (2013). The Future of Crowd Work. In *Proceedings of the 2013 Conference on Computer Supported Cooperative Work (CSCW '13)*. ACM, New York, NY, USA, 1301–1318. DOI:http://dx.doi.org/10.1145/2441776.2441923

Kittur, A, Smus, B, Khamkar, S, and Kraut, R. E. (2011). CrowdForge: Crowdsourcing Complex Work. In *Proceedings of the 24th Annual ACM Symposium on User Interface Software and Technology (UIST '11)*. ACM, New York, NY, USA, 43–52. DOI:http://dx.doi.org/10.1145/2047196.2047202

Kuhn, H. (1953). Extensive Games and the Problem of Information, Contributions to the Theory of Games II, Kuhn, HW and AW Tucker, eds. 193-216. (1953).

Kulkarni, A, Can, M, and Hartmann, B. (2012). Collaboratively Crowdsourcing Workflows with Turkomatic. In *Proceedings of the ACM 2012 Conference on Computer Supported Cooperative Work (CSCW '12)*. ACM, New York, NY, USA, 1003–1012. DOI: http://dx.doi.org/10.1145/2145204.2145354

Lasecki, W and Bigham, J. (2012)a. Self-correcting Crowds. In *CHI '12 Extended Abstracts on Human Factors in Computing Systems (CHI EA '12)*. ACM, New York, NY, USA, 2555–2560. DOI:http://dx.doi.org/10.1145/2212776.2223835

Lasecki, W, Miller, C, Sadilek, A, Abumoussa, A, Borrello, D, Kushalnagar, R, and Bigham, J. (2012). Real-time Captioning by Groups of Non-experts. In *Proceedings of the 25th Annual ACM Symposium on User Interface Software and Technology (UIST '12)*. ACM, New York, NY, USA, 23–34. DOI:http://dx.doi.org/10.1145/2380116.2380122

Lasecki, W. S and Bigham, J. P. (2012)b. Online Quality Control for Real-time Crowd Captioning. In *Proceedings of the 14th International ACM SIGACCESS Conference on Computers and Accessibility (ASSETS '12)*. ACM, New York, NY, USA, 143–150. DOI: http://dx.doi.org/10.1145/2384916.2384942

Lasecki, W. S, Gordon, M, Koutra, D, Jung, M, Dow, S, and Bigham, J. P. (2014). Glance: Rapidly Coding Behavioral Video with the Crowd. In *Proceedings of the 27th Annual ACM Symposium on User Interface Software and Technology (UIST '14)*. ACM, New York, NY, USA, 12.

Lasecki, W. S, Kamar, E, and Bohus, D. (2013)a. Conversations in the Crowd: Collecting Data for Task-Oriented Dialog Learning. In *First AAAI Conference on Human Computation and Crowdsourcing*.

Lasecki, W. S, Miller, C. D, and Bigham, J. P. (2013)b. Warping Time for More Effective Real-time Crowdsourcing. In *Proceedings of the SIGCHI Conference on Human Factors in Computing Systems (CHI '13)*. ACM, New York, NY, USA, 2033–2036. DOI: http://dx.doi.org/10.1145/2470654.2466269

Lasecki, W. S, Murray, K. I, White, S, Miller, R. C, and Bigham, J. P. (2011). Real-time Crowd Control of Existing Interfaces. In *Proceedings of the 24th Annual ACM Symposium on User Interface Software and Technology (UIST '11)*. ACM, New York, NY, USA, 23–32. DOI: http://dx.doi.org/10.1145/2047196.2047200

Lasecki, W. S, Song, Y. C, Kautz, H, and Bigham, J. P. (2013)a. Real-time Crowd Labeling for Deployable Activity Recognition. In *Proceedings of the 2013 Conference on Computer Supported Cooperative Work (CSCW '13)*. ACM, New York, NY, USA, 1203–1212. DOI: http://dx.doi.org/10.1145/2441776.2441912

Lasecki, W. S, Thiha, P, Zhong, Y, Brady, E, and Bigham, J. P. (2013)b. Answering Visual Questions with Conversational Crowd Assistants. In *Proceedings of the 15th International ACM SIGACCESS Conference on Computers and Accessibility (ASSETS '13)*. ACM, New York, NY, USA, Article 18, 8 pages. DOI: http://dx.doi.org/10.1145/2513383.2517033

Lasecki, W. S, Weingard, L, Ferguson, G, and Bigham, J. P. (2014). Finding Dependencies Between Actions Using the Crowd. In *Proceedings of the SIGCHI Conference on Human Factors in Computing Systems (CHI '14)*. ACM, New York, NY, USA, 3095–3098. DOI: http://dx.doi.org/10.1145/2556288.2557176

Lasecki, W. S, Wesley, R, Nichols, J, Kulkarni, A, Allen, J. F, and Bigham, J. P. (2013). Chorus: A Crowd-powered Conversational Assistant. In *Proceedings of the 26th Annual ACM Symposium on User Interface Software and Technology (UIST '13)*. ACM, New York, NY, USA, 151–162. DOI: http://dx.doi.org/10.1145/2501988.2502057

Lasecki, W. S, White, S, Murray, K. I, and Bigham, J. P. (2012). Crowd Memory: Learning in the Collective. In *Proceedings of Collective Intelligence*. 8.

Lee, J, Kladwang, W, Lee, M, Cantu, D, Azizyan, M, Kim, H, Limpaecher, A, Yoon, S, Treuille, A, and Das, R. (2014). RNA design rules from a massive open laboratory. *Proceedings of the National Academy of Sciences* 111, 6 (2014), 2122–2127.

Limpaecher, A, Feltman, N, Treuille, A, and Cohen, M. (2013). Real-time Drawing Assistance Through Crowdsourcing. *ACM Trans. Graph.* 32, 4, Article 54 (July 2013), 8 pages. DOI: http://dx.doi.org/10.1145/2461912.2462016

Lin, C. H, Weld, D. S, and others, . (2012). Dynamically switching between synergistic workflows for crowdsourcing. In *Workshops at the Twenty-Sixth AAAI Conference on Artificial Intelligence*.

Little, G, Chilton, L. B, Goldman, M, and Miller, R. C. (2010). TurKit: Human Computation Algorithms on Mechanical Turk. In *Proceedings of the 23Nd Annual ACM Symposium on User Interface Software and Technology (UIST '10)*. ACM, New York, NY, USA, 57–66. DOI: http://dx.doi.org/10.1145/1866029.1866040

Liu, Q, Peng, J, and Ihler, A. (2012). Variational inference for crowdsourcing. In *Advances in Neural Information Processing Systems*. 692–700.

Loparev, A, Lasecki, W. S, Murray, K. I, and Bigham, J. P. (2013). Introducing Shared Character Control to Existing Video Games. (2013).

Mao, A, Chen, Y, Gajos, K. Z, Parkes, D, Procaccia, A. D, and Zhang, H. (2012). TurkServer: Enabling Synchronous and Longitudinal Online Experiments. In *Fourth Workshop on Human Computation (HCOMP 2012)*.

Mao, A, Kamar, E, and Horvitz, E. (2013). Why Stop Now? Predicting Worker Engagement in Online Crowdsourcing. In *First AAAI Conference on Human Computation and Crowdsourcing*.

Mashhadi, A. J and Capra, L. (2011). Quality control for real-time ubiquitous crowdsourcing. In *Proceedings of the 2nd international workshop on Ubiquitous crowdsouring*. ACM, 5–8.

Mason, W and Suri, S. (2012). Conducting behavioral research on AmazonâĂŹs Mechanical Turk. *Behavior research methods* 44, 1 (2012), 1–23.

Messias, J. V, Spaan, M, and Lima, P. U. (2011). Efficient offline communication policies for factored multiagent POMDPs. In *Advances in Neural Information Processing Systems*. 1917–1925.

Michelucci, P. (2013). Synthesis and Taxonomy of Human Computation. In *Handbook of Human Computation*. Springer, 83–86.

Moldovanu, B and Sela, A. (2001). The optimal allocation of prizes in contests. *American Economic Review* (2001), 542–558.

Moldovanu, B and Sela, A. (2006). Contest architecture. *Journal of Economic Theory* 126, 1 (2006), 70–96.

Osentoski, S, Crick, C, Jay, G, and Jenkins, O. C. Crowdsourcing for closed loop control. (????).

Parkes, D. C and Procaccia, A. D. (2013). Dynamic Social Choice with Evolving Preferences. (2013), 767–773.

Proper, S and Tadepalli, P. (2009). Solving multiagent assignment markov decision processes. In *Proceedings of The 8th International Conference on Autonomous Agents and Multiagent Systems-Volume 1*. International Foundation for Autonomous Agents and Multiagent Systems, 681–688.

Quinn, A. J and Bederson, B. B. (2011). Human Computation: A Survey and Taxonomy of a Growing Field. In *Proceedings of the SIGCHI Conference on Human Factors in Computing Systems (CHI '11)*. ACM, New York, NY, USA, 1403–1412. DOI: http://dx.doi.org/10.1145/1978942.1979148

Raykar, V. C, Yu, S, Zhao, L. H, Valadez, G. H, Florin, C, Bogoni, L, and Moy, L. (2010). Learning from crowds. *The Journal of Machine Learning Research* 11 (2010), 1297–1322.

Retelny, D, Robaszkiewicz, S, To, A, Lasecki, W, Patel, J, Rahmati, N, Doshi, T, Valentine, M, and Bernstein, M. (2014). Expert Crowdsourcing with Flash Teams. In *Proceedings of the 27th Annual ACM Symposium on User Interface Software and Technology (UIST '14)*. ACM, New York, NY, USA, 12.

Rossen, B and Lok, B. (2012). A crowdsourcing method to develop virtual human conversational agents. *International Journal of Human-Computer Studies* 70, 4 (2012), 301 – 319. DOI: http://dx.doi.org/10.1016/j.ijhcs.2011.11.004

Russell, B. C, Torralba, A, Murphy, K. P, and Freeman, W. T. (2008). LabelMe: A Database and Web-Based Tool for Image Annotation. *Int. J. Comput. Vision* 77, 1-3 (May 2008), 157–173. DOI: http://dx.doi.org/10.1007/s11263-007-0090-8

Sheng, V. S, Provost, F, and Ipeirotis, P. G. (2008). Get another label? improving data quality and data mining using multiple, noisy labelers. In *Proceedings of the 14th ACM SIGKDD international conference on Knowledge discovery and data mining*. ACM, 614–622.

Singh, P, Lasecki, W. S, Barelli, P, and Bigham, J. P. (2012). *HiveMind: A Framework for Optimizing Open-Ended Responses From the Crowd*. Technical Report. URCS Technical Report.

Snow, R, O'Connor, B, Jurafsky, D, and Ng, A. Y. (2008). Cheap and Fast—but is It Good?: Evaluating Non-expert Annotations for Natural Language Tasks. In *Proceedings of the Conference on Empirical Methods in Natural Language Processing (EMNLP '08)*. Association for Computational Linguistics, Stroudsburg, PA, USA, 254–263.

Spaan, M. T and Oliehoek, F. A. (2012). Tree-Based Solution Methods for Multiagent POMDPs with Delayed Communication. In *Proc. of 24th Benelux Conference on Artificial Intelligence*. 319–320.

Surowiecki, J. (2005). *The wisdom of crowds*. Random House LLC.

Van Kleek, M. G, Bernstein, M, Panovich, K, Vargas, G. G, Karger, D. R, and Schraefel, M. (2009). Note to Self: Examining Personal Information Keeping in a Lightweight Note-taking Tool. In *Proceedings of the SIGCHI Conference on Human Factors in Computing Systems (CHI '09)*. ACM, New York, NY, USA, 1477–1480. DOI: http://dx.doi.org/10.1145/1518701.1518924

von Ahn, L. (2005). *Human Computation*. Ph.D. Dissertation. Carnegie Mellon University, Pittsburgh, PA.

von Ahn, L and Dabbish, L. (2004). Labeling Images with a Computer Game. In *Proceedings of the SIGCHI Conference on Human Factors in Computing Systems (CHI '04)*. ACM, New York, NY, USA, 319–326. DOI: http://dx.doi.org/10.1145/985692.985733

Wikipedia (2014). Words per minute. (2014). http://en.wikipedia.org/wiki/Words_per_minute Accessed: 2014-06-05.

Zhang, H, Law, E, Miller, R, Gajos, K, Parkes, D, and Horvitz, E. (2012). Human Computation Tasks with Global Constraints. In *Proceedings of the SIGCHI Conference on Human Factors in Computing Systems (CHI '12)*. ACM, New York, NY, USA, 217–226. DOI: http://dx.doi.org/10.1145/2207676.2207708

Zhou, D, Basu, S, Mao, Y, and Platt, J. C. (2012). Learning from the wisdom of crowds by minimax entropy. In *Advances in Neural Information Processing Systems*. 2195–2203.

Human Computation (2014) 1:1:95-99
© 2014, Bateson. CC-BY-3.0
ISSN: 2330-8001, DOI: 10.15346/hc.v1i1.6

Learning to Count

MARY CATHERINE BATESON, Cultural Anthropologist

[Editor's note: this is a reworking of Mary Catherine Bateson's original piece on "Making a Difference", which was the foreword to the Springer Handbook of Human Computation (2013).]

One of the mysteries of our species, now approaching eight billion persons, is the very limited way in which the intelligence of this vast number of individuals can be said to add up or combine. On the one hand, every society has a body of knowledge that is passed on from generation to generation, with frequent borrowing from neighbors and gradual enrichment with new information and new understanding. Human history can even be seen as the steady process of discovering ways to store and pass on information and ideas. On the other hand, differing ideas lead to failures to cooperate and misunderstandings, complicated by the emphasis on competition and the notion that knowledge and ideas are a form of property. All these wonderful brains operate in various degrees of separation, so that neither nations nor the species as a whole has the benefit of their combined power. Do we – that is to say, all members of the human species together – have the intelligence to order our living in such a way as to preserve the viability of our planet? Surely. Do we have the capacity to integrate that intelligence, to mobilize its combined potential toward that goal, and to act on it? Not yet.

The question has been around for a long time. The myth of the Tower of Babel in Genesis is evidence of long ago puzzlement about the human difficulty in coordinating creative action with inadequate communication. In our day, when massive computing power is wanted, the solution tends to be parallel processing – multiple separate computers linked together. Early census collections were undoubtedly collated by groups of clerks toiling side by side, perhaps a little like the "boiler rooms" in which multiple salesmen make cold calls to potential investors. But setting computers or clerks side by side, while it makes it possible to crunch huge quantities of data, does not offer a model for benefiting from the diversity of different human minds complementing each other. Voting is a form of human computation that requires the same kind of input from multiple individuals in order to make a choice. Polling collates simple answers to the same question, shorn of reflection. Neither of these procedures generates new ideas.

The field of human computation, then, has two faces. On the one hand, there is the aggregation of the effort of many different persons doing the same task or making similar inputs from different places, perhaps in-putting data about observations of threatened species or meteorological phenomena. On the other hand, there is the potential for the integration of multiple different kinds of input coming from diverse individuals to produce new and creative possibilities. This is ideally done in conversation where the participants are stimulated by their diverse points of view, aiming to discover new alternatives or to arrive at a consensus, to become "of one mind." When the numbers involved make conversation awkward, the integration process can be assisted by technology.

Thus, there have been in recent decades a wide variety of proposed methods for facilitating productive conversation that may then be collated electronically. Even when the inputs are similar in kind, there is the possibility that the aggregation of multiple responses can be an important step toward solving a fundamental ethical problem in human society, namely the increasingly widespread conviction that "nothing I can do will make any difference." Kant's Categorical Imperative[1] was an attempt to solve the problem by eliminating the question of scale and proposing that an action be evaluated as if it were universal, but this has not proved particularly effective in ever larger populations. The problem of taking responsibility for individual and local actions is most severe at the global level. Thus, for instance, individuals have difficulty believing that leaving an extra electric light burning in their suburban backyard is connected to the likelihood of lethal storms thousands of miles away. Exactly the same kind of reasoning discourages voters from going to the polls for local elections. How will people learn that what they do "counts"? By counting. Similarly, the endless series of petitions posted on the Internet and the more and more frequent demands to "rate our service," are intended to give people the sense of contributing to common goals. The Vatican recently invited bishops to poll the faithful, and many responded and hope that their opinions will be heard and integrated in decision-making – that their words would really count.

We badly need models of interdependence and connectivity that will convey to those who work with them the conviction that individual voices and actions do indeed count, a message conveyed through many different modalities. It is impossible to cultivate a sense of responsibility without such a conviction. It also seems probable that ethics is meaningless without an understanding of the different kinds of causation, including indirect, multiple, and circular causation. Human computation for socially useful goals will depend on giving individuals a sense of agency – a sense that they indeed can make a difference – and a sense of how different kinds of action interact for good or for ill.

---

[1] "There is, therefore, only one categorical imperative. It is: Act only according to that maxim by which you can at the same time will that it should become universal law." Immanuel Kant, Foundations of the Metaphysics of Morals, trans. Lewis W. Beck, ed. Robert P. Wolff, section 2, p. 44.

Agency has been the central issue for the development of patient activism, both spurring research and leading to better prevention and intelligent compliance, so enrolling patients as active collaborators in research has been an important new model for citizen science. One of the earliest examples of citizen science was the St. Louis baby tooth collection organized by Barry Commoner, in which scientists "took over the tooth fairy" (Bateson, 1972) to demonstrate the dangers of nuclear testing in the atmosphere. The demonstration that Strontium 90 was being transferred in mothers' milk was a significant element in the banning of atmospheric testing, but so no doubt was the engagement it evoked in the parents.

There is a long history, going back to the Greeks and Romans, of attempting to use voting (an early form of human computation), with various modifications, to create a sense of agency that supports responsibility, and some of the hazards are known. It is not enough for bishops to hand out (or post) questionnaires: people must be convinced that they read them. Experience suggests that plebiscites are easily manipulated by autocrats (as in the rise of fascism), so that it makes more sense to vote for individuals who are then able to deliberate together about issues and act systemically as surrogate decision makers in a second round of voting than it does to decide policy by majority popular vote, thus combining computation with the integrative potential of conversation. Other variations such as proportional representation also attempt to avoid the dangers of simple majority rule. Voter initiatives may appear to increase democracy but when overused may lock in dysfunctional policies. Ventures into human computation need to be evaluated with an eye to whether the inclusion of multiple inputs will lead to better and potentially wiser decisions. And all such ventures should contain provision for feedback. A friend tells me that she spends an hour every morning signing petitions (on the Internet). Real feedback is rare – the usual, not very reinforcing reward is generally a demand for donations – but Walgreen's did apparently give up the decision to move its corporate address out of the US in response to an Internet petition, and this news went out to all signers.

A central promise of human computation, already partially realized, is the possibility of creating an awareness of the vast number of decisions we all make every day, including the decision involved in where attention is focused from minute to minute (Jackson, 2009), along with information about the aggregate effect of those decisions and how they are shifting. Does my driving behavior or the way I handle my trash make me part of a great and growing expression of commitment to preserve the planet? Is this a case where positive feedback can be truly regenerative?

Information about new ideas and emerging patterns needs to be accessible and individual voices need to be audible, but human computation may run the risk of simply reinforcing existing trends, which may be negative, by facilitating conformity. The popularity of SUVs and violent movies and games tends to be self-reinforcing and the most popular restaurant in town may not be the most pleasant place to go on a Saturday night. Simply waiting to see what "goes viral" on YouTube or Twitter is not sufficient. A noteworthy variation on regenerative feedback, however, is Kickstarter.com, which works like a chain letter to raise funds for nonprofit projects.

Another significant trend related to Human Computation is the effort to create interactive contexts for the expression of greater diversity of knowledge and imagination. Interdisciplinary conferences (such as the Macy conferences on Cybernetics and on Group Process in the 1940s (Heims, 1991)) can be seen as an example of taking a group of individuals and turning them into a thinking system, a kind of superorganism (M.C.Bateson, 1972). With the decline in support for exploratory interdisciplinary work, there has been a rise in designs for interactive process in facilitating meetings, such as America Speaks, the 21st century Town Hall Meeting format devised by Carolyn Lukensmeyer (Lykensmeyer, 2007) and Laura Chasin's Public Conversations Project (Herzig and Chasin, 2006), as well as research on conflict resolution and mediation (Fisher, 1991), (simultaneously alas with the steady increase in what Deborah Tannen calls the Argument Culture, in which issues are approached antagonistically (Tannen, 1998). Innovative techniques like brainstorming or open space exercises can themselves be regarded as technologies of human computation.

Human beings change in response to their habitual interactions, and there is already concern about possible deleterious effects of electronic communication, which will surely play a major role in human computation as we move forward. Much of human computation depends on persuading large numbers of individuals, acting separately, to contribute personal information, which is then combined, both processes facilitated by electronic technology. But it is important to notice that the implicit message of such an operation is membership in a larger whole. Any living system processes quantities of material and information, in ways that affect the state of that system and other systems to which it is connected, and attending to such processes potentially creates a sense of unity and an awareness of the reality of interdependence.

Emphasis on individualism has weakened for many people an understanding of what it means to be a part of some larger system, like the biosphere of this planet. We know today that our entire planet can be looked at as a living system (Lovelock, 1995) with some capacity for self-regulation, and that the circulation of water and atmospheric gases is such that disruption or pollution in one place on the planet has measurable effects elsewhere. Indeed, earth systems are far more closely integrated than the present human capacity to respond to them, even in the preparation for and response to major disasters. The American emphasis on individual autonomy is a product of the circumstances under which Europeans settled the North American continent, but it is descriptively inaccurate for the human condition and inhibits effective cooperation in problem solving and humanitarian relief as we experience and attempt to mitigate the global effects of climate disruption. Arguably, then, if increased reliance on human computation shifts attitudes away from the fetish of individual autonomy and teaches us, by implication, to recognize that we are connected parts of a larger whole, this is a goal to be pursued. Perhaps too, the awareness of inescapably "making a difference," for better or for worse, by our individual choices, will come to be seen as an essential aspect of human dignity as we learn that all of our choices count.

REFERENCES

Bateson, M. C. (1972). *Our Own Metaphor: A Personal Account of a Conference on the Effects of Conscious Purpose on Human Adaptation*. New York: Alfred A. Knopf.

Bateson, M. C. (2013). Foreword: Making a Difference. In P. Michelucci (Ed.), *Handbook of Human Computation* (pp. 83–86). Springer New York.

Fisher, R., Ury, W. L., & Patton, B. (1991). *Getting to Yes: Negotiating Agreement Without Giving In* (Revised edition.). New York, N.Y: Penguin Books.

Heims, S. J. (1991). *The Cybernetics Group*. Cambridge, Mass: The MIT Press.

Herzig, M. and Chasin, L. Fostering dialogue across divides: A nuts and bolts guide from the Public Conversations Project. Watertown, MA: The Public Conversations Project, 2006. Available at www.publicconversations.org

Holman, P., Devane, T., & Cady, S. (2007). *The Change Handbook: The Definitive Resource on Today's Best Methods for Engaging Whole Systems* (2nd edition.). San Francisco: Berrett-Koehler Publishers.

Jackson, M. (2009). *Distracted: the erosion of attention and the coming Dark Age*. Amherst, N.Y.: Prometheus Books.

Lovelock, J. (1995). *A New Look at Life on Earth*. Oxford and New York: Oxford University Press.

Tannen, D. (1998). *The Argument Culture: Moving from Debate to Dialogue* (1st edition.). New York: Random House.

www.ingramcontent.com/pod-product-compliance
Lightning Source LLC
Chambersburg PA
CBHW050851180526
45159CB00007B/2644

REFERENCES

Bateson, M. C. (1972). *Our Own Metaphor: A Personal Account of a Conference on the Effects of Conscious Purpose on Human Adaptation.* New York: Alfred A. Knopf.

Bateson, M. C. (2013). Foreword: Making a Difference. In P. Michelucci (Ed.), *Handbook of Human Computation* (pp. 83–86). Springer New York.

Fisher, R., Ury, W. L., & Patton, B. (1991). *Getting to Yes: Negotiating Agreement Without Giving In* (Revised edition.). New York, N.Y: Penguin Books.

Heims, S. J. (1991). *The Cybernetics Group.* Cambridge, Mass: The MIT Press.

Herzig, M. and Chasin, L. Fostering dialogue across divides: A nuts and bolts guide from the Public Conversations Project. Watertown, MA: The Public Conversations Project, 2006. Available at www.publicconversations.org

Holman, P., Devane, T., & Cady, S. (2007). *The Change Handbook: The Definitive Resource on Today's Best Methods for Engaging Whole Systems* (2nd edition.). San Francisco: Berrett-Koehler Publishers.

Jackson, M. (2009). *Distracted: the erosion of attention and the coming Dark Age.* Amherst, N.Y.: Prometheus Books.

Lovelock, J. (1995). *A New Look at Life on Earth.* Oxford and New York: Oxford University Press.

Tannen, D. (1998). *The Argument Culture: Moving from Debate to Dialogue* (1st edition.). New York: Random House.

www.ingramcontent.com/pod-product-compliance
Lightning Source LLC
Chambersburg PA
CBHW050851180526
45159CB00007B/2644